Assessment of Risks Related to Wellbore Safety in Deepwater Drilling

Yuqiang Xu

Assessment of Risks Related to Wellbore Safety in Deepwater Drilling

中国石油大学出版社
CHINA UNIVERSITY OF PETROLEUM PRESS

Springer

Yuqiang Xu
China University of Petroleum (East China)
Qingdao, China

ISBN 978-981-95-1667-4 ISBN 978-981-95-1668-1 (eBook)
https://doi.org/10.1007/978-981-95-1668-1

Jointly published with China University of Petroleum Press
The print edition is not for sale in China (Mainland). Customers from China (Mainland) please order the print book from: China University of Petroleum Press.

This Springer imprint is published by the registered company Springer Nature Singapore Pte Ltd.
The registered company address is: 152 Beach Road, #21-01/04 Gateway East, Singapore 189721, Singapore

Preface

Deepwater oil and gas resources are the main areas of energy replacement that have attracted global attention and are being explored and developed in the twenty-first century. It is reported that more than half of the major global marine oil and gas discoveries were concentrated in deep waters. Take China as an example, the oil and gas resources in the South China Sea of China are also very abundant, with 70% of them. In 2014, China drilled the first deepwater exploration well in the Lingshui 17-2 gas field and discovered China's first 100-billion-cubic-meter self-operated deepwater gas field—the "DeepSea No. 1" gas field. As of 2025, it has cumulatively produced over 10 billion cubic meters of natural gas and over 1 million cubic meters of condensate oil. The "DeepSea No. 1" gas field has a maximum operational water depth exceeding 1500 m, a maximum well depth reaching over 5000 m, a stratigraphic maximum temperature of 138 °C, and a stratigraphic maximum pressure exceeding 69 MPa.

Deepwater drilling engineering is a necessary step and preliminary guarantee for the exploration and development of deep-sea oil and gas resources. Due to factors such as great water depth, long riser pipelines, low seabed temperature, shallow geological disasters, a narrow safe drilling fluid density window, and the need for special drilling processes and equipment, many problems and potential risks have been brought to the design and construction of deepwater drilling engineering. Since the exploration and development of marine oil and gas, especially with the development of deep-sea oil and gas resources, quite a number of major accidents have occurred due to insufficient estimation of potential risks. For example, in the oil spill accident in the Gulf of Mexico, potential risks such as natural gas hydrates and wellbore integrity were ignored in the design and construction of deepwater drilling and completion wells, ultimately leading to the explosion and sinking of the drilling platform, causing immeasurable losses to human lives, property, and the environment. It can be seen that while deepwater drilling engineering has high investment and high returns, it also has high risks. How to predict risks in the design of deepwater drilling engineering, control and avoid risks during the design and construction process, and propose specific evaluations and solutions for potential risks have become hot topics and important tasks in today's deepwater drilling engineering.

According to the characteristics of deepwater drilling engineering, this book divides its wellbore profile into two parts: the section from the sea level to the subsea wellhead and the part below the seabed mud line. In deepwater drilling, the upper part of the subsea wellhead is connected to the blowout preventer group and a relatively long riser pipeline, and the lower part is connected to a series of casings, resulting in the subsea wellhead being subjected to the stress and gravity from both the upper and lower parts simultaneously, and being in a complex stress environment with the risk of instability. Although the part below the seabed mud line in deepwater drilling is not very different from land drilling, there are still challenges such as greater water depth, low seabed temperature, a narrow safe density window, shallow geological disasters, natural gas hydrates, and difficulties in determining drilling geological parameters, which greatly increase the risks of the drilling engineering in this part. Based on systematically expounding the basic theories and methods such as the mechanical stability of the subsea wellhead in deepwater drilling, the calculation of the wellbore pressure field during gas invasion, the prediction of the hydrate formation area, and the design method of the wellbore structure, this book analyzes the risks of problems such as wellhead instability, gas invasion and blowout, natural gas hydrates in the wellbore, and wellbore structure design, which have relatively high risks during the deepwater drilling process. It also uses methods such as probability statistics to explore the characteristics and influence laws of the uncertain parameters and empirical coefficients, and establishes a risk assessment method related to the safety of the deepwater drilling wellhead and wellbore considering the uncertain parameters. Thus, it quantitatively evaluates and analyzes the risk categories and probabilities of many risks and gives corresponding risk avoidance suggestions or solutions, forming a systematic theory and method for the safety risk assessment of deepwater drilling wellbores. This will be of great significance for the prevention and control of risks in deepwater drilling, the optimization of the design and construction scheme of deepwater drilling, the improvement of the lifespan and investment returns of deepwater oil and gas wells, and the promotion of the development of deepwater drilling technology.

This book mainly introduces basic theoretical methods and case studies. With the on-site practice and iterative development of relevant theoretical technologies, the consideration of factors has become more comprehensive, and the research depth has gradually increased. For example, the riser gas invasion monitoring method proposed in this book has formed relatively mature experimental and numerical simulation research methods in subsequent studies, and even field tests have been carried out. The author also proposes methods to further move the monitoring downward to the near-bit area and use acoustic-electric multi-physical field coupling monitoring for complex overflow. It is hoped that the content of this book can inspire peers, scholars, and engineers.

Qingdao, China Yuqiang Xu

Competing Interests The author has no competing interests to declare that are relevant to the content of this manuscript.

Contents

Chapter 1
Risk Assessment Methodology for Mechanical Stability of Subsea Wellheads in Deepwater Drilling

Abstract The primary distinction between deepwater drilling and shallow-water drilling lies in the greater water depth, necessitating the use of long risers to ensure drilling operations. Additionally, as water depth increases, the mechanical properties of shallow seabed soils change. Under the influence of factors such as drilling vessel drift and ocean current impacts, subsea wellheads face instability risks. Existing mechanical stability calculation models for wellheads often employ deterministic parameters. However, the complex environment of deepwater drilling introduces numerous uncertainties, reducing the accuracy and reliability of these models. Therefore, based on the study of mechanical stability calculation models for deepwater subsea wellheads, it is essential to analyze sensitive and uncertain factors and establish a risk assessment methodology that accounts for uncertainties. This approach provides technical support for quantitatively evaluating and mitigating instability risks in deepwater subsea wellheads.

1.1 Mechanical Stability Calculation Model for Deepwater Subsea Wellheads

The riser-seabed wellhead-conductor system serves as a critical safety barrier to isolate external environmental complexities, support subsequent tubulars, and ensure smooth deepwater drilling operations. However, this system faces safety risks due to drilling vessel drift, ocean current impacts, and shallow geological hazards. Within this system, the subsea wellhead is subjected to dual forces from the riser and conductor, making it more prone to instability. Once destabilized, catastrophic consequences may occur. Consequently, domestic and international scholars have conducted extensive research on the mechanical stability of deepwater subsea wellheads.

Instability in deepwater subsea wellheads primarily manifests in two forms: wellhead tilting and subsidence.

Y. Xu, *Assessment of Risks Related to Wellbore Safety in Deepwater Drilling*, https://doi.org/10.1007/978-981-95-1668 1_1

(1) Wellhead Tilting: Caused by riser deformation under drilling vessel drift or ocean current impacts, leading to lateral displacement of the wellhead. This typically occurs during post-surface-casing cementing operations.
(2) Wellhead Subsidence: Results from insufficient conductor design depth, where the conductor's bearing capacity after installation cannot support the weight of the wellhead assembly and subsequent casings. This usually occurs before surface casing cementing.

Thus, the mechanical stability of subsea wellheads can be studied from two perspectives: lateral mechanical stability and longitudinal mechanical stability.

1.1.1 Lateral Mechanical Stability Calculation Model

Analyzing the lateral mechanical stability of a subsea wellhead requires comprehensive consideration of environmental loads, drilling vessel drift, riser top tension, shallow seabed soil properties, and conductor depth. Therefore, the riser, wellhead, and conductor must be treated as an integrated system for holistic mechanical analysis. Over years of development and refinement, coupled mechanical analysis models for the riser-wellhead-conductor system have achieved sufficient accuracy for engineering applications. This book employs the method established by Su Kanhua to analyze the lateral mechanical stability of deepwater subsea wellheads.

The force analysis diagram of the riser-subsea wellhead-conductor system is shown in Fig. 1.1.

The force diagram of the riser-wellhead-conductor system (Fig. 1.1) reveals that the wellhead is subjected to complex forces, including lateral and vertical loads, as well as non-uniformly distributed pressures. The upper section is influenced by the riser's bottom flex/spherical joint, the lower section by the conductor and suspended casing, and lateral forces from ocean currents. Key considerations include: Riser deformation due to drilling platform drift and ocean currents, generating lateral forces on the wellhead. Conductor deformation influenced by seabed soil resistance, requiring soil properties to be integrated into the mechanical model. To balance accuracy and simplicity, the following assumptions are made:

(1) The riser is modeled as a homogeneous, linear-elastic circular pipe, ignoring the impact of drill strings and choke lines on bending stiffness.
(2) Only small deformations occur in the riser under external forces.
(3) The riser bottom and BOP are treated as rigid bodies unaffected by ocean currents.

1.1.1.1 Governing Equations for Lateral Mechanical Stability

Based on the force analysis in Fig. 1.1 and the aforementioned assumptions, the differential equations governing the lateral mechanical stability of the riser-wellhead-conductor system are derived:

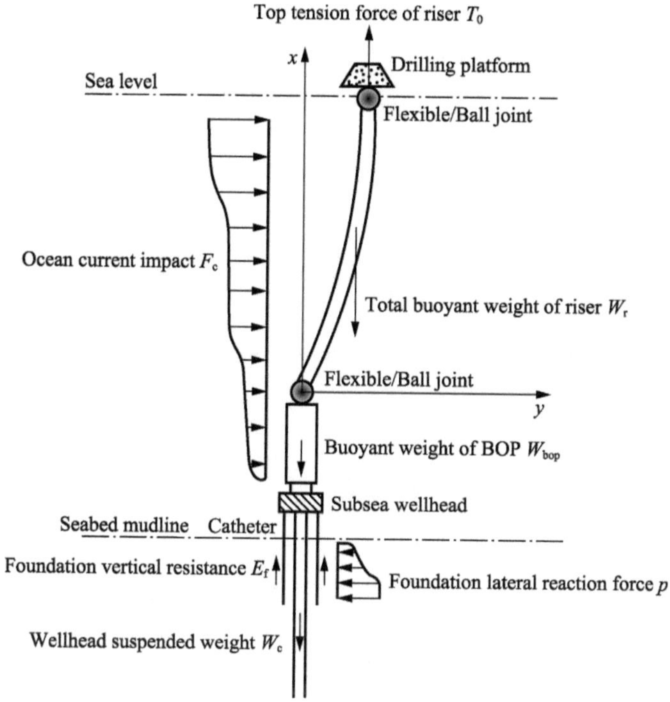

Fig. 1.1 Riser-subsea wellhead-conductor system force analysis diagram

$$\begin{cases} E_r I_r \frac{d^4y}{dx^4} - T(x)\frac{d^2y}{dx^2} - w_r \frac{dy}{dx} = F_c(x), 0 \le x \le L_r \text{ (a)} \\ \frac{d^2}{dx^2}\left(K(x)\frac{d^2y}{dx^2}\right) + \frac{d}{dx}\left(N(x)\frac{dy}{dx}\right) + D_s(x)p(x, y) = 0, -L_{bop} \ge x \ge -\left(L_{bop} + L_{sc}\right) \text{ (b)} \end{cases} \tag{1.1}$$

In the formula, $E_r I_r$ is the flexural rigidity of the riser, kN·m^2; $T(x)$ is the effective tension of the riser, kN; w_r is the weight per unit length of the riser; $F_c(x)$ is the lateral ocean current impact force per unit length of the riser, kN; $K(x)$ is the equivalent flexural rigidity of the combined string of the conductor, surface casing and cement sheath, kN·m^2; $N(x)$ is the axial force of the string composed of the conductor, surface casing and cement sheath, kN; $D_s(x)$ is the outer diameter of the conductor and surface casing, m; $p(x, y)$ is the foundation reaction force acting on the unit area of the conductor, kPa; L_{bop} is the height of the Blowout Preventer (BOP), m; L_{sc} is the length of the combined string of the conductor and surface casing, m.

Formulas for parameters in the differential equations:

(1) Riser weight per unit length w_r:

$$W_r = \frac{\pi}{4}\left(D_r^2 - d_r^2\right)\rho_s g + \frac{\pi}{4} d_r^2 \rho_m g - \frac{\pi}{4} D_r^2 \rho_l g - B \tag{1.2}$$

(2) Effective riser tension $T(x)$:

$$T(x) = T_0 - \int_x^{L_r} W_r \mathrm{d}x \tag{1.3}$$

(3) Lateral ocean current force per unit length $F_c(x)$:

$$F_c(x) = \frac{1}{2} C_D \rho_l D_r \left(u_t \left(\frac{x}{L_r} \right)^{1/7} + u_w \frac{x}{L_r} \right)^2 \tag{1.4}$$

(4) Equivalent bending stiffness of conductor assembly:

$$K = \begin{cases} E_{stl}(I_{so} + I_{si}) \\ E_{stl} I_{si} + 0.8 E_c I_c \\ E_{stl}(I_{so} + I_{si}) + 0.6 E_c I_c \end{cases} \tag{1.5}$$

In the formula, $E_{stl}(I_{so} + I_{si})$ is the conductor pipe-surface casing combined string; $E_{stl} I_{si} + 0.8 E_c I_c$ is the surface casing-cement sheath combined string; $E_{stl}(I_{so} + I_{si}) +$ $0.6 E_c I_c$ is the conductor pipe-surface casing-cement sheath combined string.

(5) Axial force of conductor assembly $N(x)$:

$$N(x) = \begin{cases} N_t - W_c(x) \cdot x, x \geq -x_{ml} \\ N_t - W_c(x) \cdot x + F_f(x) \cdot (x_{ml} - x), x < -x_{ml} \end{cases} \tag{1.6}$$

(6) Foundation reaction per unit area $p(x, y)$:

Matlock H. and Reese L.C. established the *p-y* curve in the 1970s through experiments, which is adopted here for $p(x, y)$ calculations (details omitted).

In the above equations: T_0 is the top tension of the riser, kN; D_r is the outer diameter of the riser, m; d_r is the inner diameter of the riser, m; ρ_s is the density of the riser, kg/m^3; ρ_m is the density of the drilling fluid, kg/m^3; ρ_l is the density of seawater, kg/m^3; B is the buoyancy generated by the buoyancy section per unit length, kN; C_D is the drag coefficient; μ_t and μ_w are the tidal current velocity and wind—driven current velocity at the water surface respectively, m/s; E_{st1} is the elastic modulus of the steel used for the conductor and surface casing, kPa; E_c is the elastic modulus of the cement sheath, kPa; I_{so}, I_{si}, and I_c are the sectional moments of inertia of the conductor, surface casing, and cement sheath respectively, m; N_t is the vertical force acting on the wellhead, kN; W_c is the weight per unit length of the combined string, kN; $F_f(x)$ is the frictional force of the soil on the outer wall of the conductor per unit length, kN; x_{ml} is the height of the combined string above the seabed mudline, m.

1.1.1.2 Continuity and Boundary Conditions

As analyzed in Fig. 1.1, the BOP, wellhead, conductor, and surface casing form an integrated system. The riser connects to the BOP via a flex/spherical joint at its lower end. Continuity conditions are established as follows:

$$\begin{cases} y_r = y_c + L_{bop} \cdot sin(\theta_c) \\ y_{re}\big|_{x=0} = y_r,\ y_{cs}\big|_{x=-L_{bop}} = y_c \end{cases} \tag{1.7}$$

For the riser, its upper boundary is the flexible/spherical joint at the top end of the riser, and the lower boundary is the flexible/spherical joint at the lower end of the riser. The boundary conditions can be calculated using the bending moment and lateral displacement. Since the Blowout Preventer (BOP) was assumed to be a rigid body previously, for the whole structure composed of the BOP, subsea wellhead, conductor, and surface casing, the upper boundary can be counted from the subsea wellhead, and the lower boundary is the bottom of the conductor and surface casing. The boundary conditions can be calculated using the bending moment and cross-sectional shear force:

$$\begin{cases} M\,|x = L_r = K_{ru}\theta_{ru}, & y_{re}\big|_{x=L_r} = S_0 \\ M\,|x = 0 = K_{rd}\theta_{rd}, & y_{re}\big|_{x=0} = y_r \\ M\,|x = -L_{bop} = M_0, & Q\big|_{x=-L_{bop}} = -H_0 \\ M\,|x = -\left(L_{bop} + L_{sc}\right) = 0, & Q\big|_{x=-\left(L_{bop}+L_{sc}\right)} = 0 \end{cases} \tag{1.8}$$

In the formula, θ_c is the rotation angle at the subsea wellhead, rad; y_{re} and y_{cs} are the lateral offsets of the riser and the combined string respectively, m; K_{ru} and K_{rd} are the rotational stiffnesses of the flexible/spherical joints at the top and lower ends of the riser respectively, kN·m/rad; θ_{ru} and θ_{rd} are the rotation angles of the flexible/spherical joints at the top and lower ends of the riser respectively, rad; S_0 is the lateral displacement of the drilling platform, m; y_r and y_c are the lateral offsets at the flexible/spherical joint at the lower end of the riser and the subsea wellhead respectively, m; M_0 is the bending moment at the subsea wellhead, kN·m; H_0 is the lateral force at the subsea wellhead, in kN; Q is the shear force at the cross-section of the combined string, kN.

1.1.1.3 Discretization of Governing Equations

The finite difference method is used to discretize the control equations and boundary conditions. The riser is divided into n equal parts starting from the lower end, and the length of each element is h. The combined string is divided into n equal parts starting from the top end, and the length of each unit is also h. Then the control equation set (1.1) can be transformed into the following difference form:

$$\begin{cases} 2E_rI_ry_{re}(i+2)+\left[-8E_rI_r-2T(i)h^2-W(i)h^3\right]y_{re}(i+1)+\left[12E_rI_r+4T(i)h^2\right]y_{re}(i) \\ +\left[-8E_rI_r-2T(i)h^2+W(i)h^3\right]y_{re}(i-1)+2E_rI_ry_{re}(i-2)=2F_ch^4 \quad \text{(a)} \\ K(i+1)y_{cs}(i+2)+\left[-2K(i+1)-2K(i)+N(i)h^2\right]y_{cs}(i+1)+[K(i+1)+4K(i) \\ +K(i-1)-2N(i)h^2+D_s(i)E_s(i)h^4]y_{cs}(i)+\left[-2K(i)-2K(i-1)+N(i)h^2\right]y_{cs}(i-1) \\ +K(i-1)y_{cs}(i-2)=0 \quad \text{(b)} \end{cases} \tag{1.9}$$

The continuity conditions and boundary conditions can be transformed into the following difference forms:

$$\begin{cases} y_{re}(n+1)=\frac{4}{\left(2-\frac{K_{ru}h}{E_rI_r}\right)}y_{re}(n)-\frac{\left(2+\frac{K_{ru}h}{E_rI_r}\right)}{\left(2-\frac{K_{ru}h}{E_rI_r}\right)}y_{re}(n-1),\ y_{re}(n)=S_0 \quad \text{(a)} \\ y_{re}(-1)=\frac{\left(2-\frac{K_{rd}h}{E_rI_r}\right)}{\left(-2-\frac{K_{rd}h}{E_rI_r}\right)}y_{re}(1)+\frac{4}{\left(2-\frac{K_{rd}h}{E_rI_r}\right)}y_{re}(0),\ y_{re}(0)=y_r \quad \text{(b)} \\ y_{cs}(1)-2y_{cs}(0)+y_{cs}(-1)=\frac{h^2}{K_0}M_0 \\ K(1)y_{cs}(2)-\left(2K(1)-N(0)h^2\right)y_{cs}(1)+(K(1)-K(-1)y_{cs}(0)+ \\ \left(2K(-1)-N(0)h^2\right)y_{cs}(-1)-K(-1)y_{cs}(-2)=2h^3H_0 \quad \text{(c)} \\ y_{cs}(n+1)-2y_{cs}(n)+y_{cs}(n-1)=0 \\ K(n+1)y_{cs}(n+2)-\left(2K(n+1)-N(n)h^2\right)y_{cs}(n+1)-K(n-1)y_{cs}(n)+ \\ \left(2K(n-1)-N(n)h^2\right)y_{cs}(n-1)-K(n-1)y_{cs}(n-2)=0 \quad \text{(d)} \end{cases} \tag{1.10}$$

In the formula, E_s is the secant modulus of the foundation reaction force at the position x, which is determined by the *p*-*y* curve, $E_s=p/y$.

1.1.1.4 Solution Procedure

(1) Assume that the offset $y^*=y_r=0$ at the flexible/spherical joint at the lower end of the riser. Solve Eqs. (1.9a), (1.10a) and (1.10b) to obtain the tension and offset angle at the flexible/spherical joint at the lower end of the riser.
(2) According to the mechanical analysis of the subsea wellhead in Fig. 1.1, calculate the bending moment M_0, vertical force N_0 and lateral force H_0 at the subsea wellhead.
(3) Substitute M_0 and H_0 into Eq. (1.10c), and jointly solve Eqs. (1.9b), (1.10c) and (1.10d) to calculate the lateral displacement y_c and offset angle θ_c at the subsea wellhead.
(4) Substitute y_c and θ_c into Eq. (1.7), calculate the displacement y_r at the flexible/spherical joint at the lower end of the riser and update the value of $y^*=y_r$ in step (1). Repeat steps (1) to (4). The loop termination condition is: $|y^*-y_r|<\varepsilon$.
(5) Record the lateral displacement $y_{cs}(i)$ at each node, and calculate the offset angle, bending moment, shear force and foundation reaction force at each node according to Eq. (1.11):

$$\begin{cases} \theta(i) = -\frac{y_{cs}(i+1) - y_{cs}(i-1)}{2h} \\ M(i) = -\frac{K(i)(y_{cs}(i+1) - 2y_{cs}(i) + y_{cs}(i-1))}{h^2} \\ Q(i) = -\frac{K(i)(y_{cs}(i+2) - 2y_{cs}(i+1) + 2y_{cs}(i-1) - y_{cs}(i-2))}{2h^3} \\ p(i) = E_s(i) y_{cs}(i) \end{cases} \tag{1.11}$$

1.1.2 Longitudinal Mechanical Stability Calculation Model

As analyzed earlier, wellhead subsidence primarily occurs after conductor installation but before surface casing cementing. Post-cementing, the conductor-surface casing-cement sheath assembly can withstand longitudinal loads, making subsidence rare. Therefore, longitudinal stability focuses on the conductor-surface casing assembly during surface casing suspension (Fig. 1.2).

The longitudinal stability of the wellhead depends entirely on the conductor's bearing capacity. Subsidence occurs when the combined load from the wellhead, jetting tools, conductor, and surface casing exceeds this capacity. Current field analyses often use conservative methods, ignoring time-dependent soil strength gains to ensure safety. However, integrating time effects into bearing capacity calculations can reduce deepwater drilling costs. The following methodology is widely accepted:

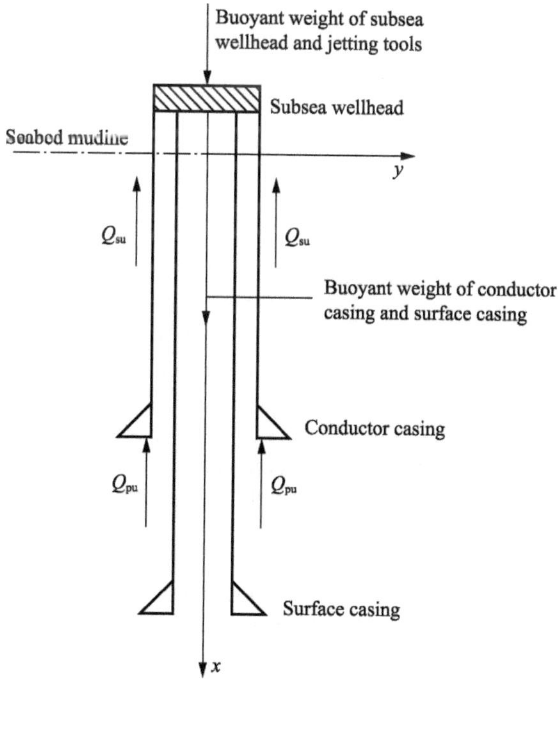

Fig. 1.2 Force analysis diagram of conductor in place with hanging surface casing

(1) Initial Conductor Bearing Capacity:

$$Q_0 = RB_{\mathrm{fw}}[(W_{\mathrm{con}} + W_{\mathrm{col}})x + W_{\mathrm{lh}} + W_{\mathrm{tool}}] \tag{1.12}$$

(2) Total Load After Surface Casing Suspension:

$$Q_{\mathrm{w}} = B_{\mathrm{fd}}(W_{\mathrm{con}}x + W_{\mathrm{scr}}L_{\mathrm{sc}}) + B_{\mathrm{fw}}(W_{\mathrm{lh}} + W_{\mathrm{hh}}) \tag{1.13}$$

(3) Ultimate Conductor Bearing Capacity:

$$Q_{\mathrm{u}} = \pi D_{\mathrm{c}} \sum_{i=0}^{n} q_{\mathrm{sui}} L_i + \frac{\pi}{4}(D_{\mathrm{c}}^2 - d_{\mathrm{c}}^2) q_{\mathrm{pu}} \tag{1.14}$$

(4) Real-Time Bearing Capacity:

$$Q_{\mathrm{t}} = Q_0 + \alpha_{\mathrm{b}}(1 + logt)(Q_{\mathrm{u}} - Q_0) \tag{1.15}$$

(5) Real-Time Bearing Capacity After Recovery Time:

Combining Eqs. (1.12), (1.14), and (1.15):

$$\begin{aligned} Q_{\mathrm{t}}^* = {} & B_{\mathrm{fw}}[(W_{\mathrm{con}} + W_{\mathrm{col}})x + W_{\mathrm{lh}} + W_{\mathrm{tool}}] + \alpha_{\mathrm{b}}(1 + logt) \geq \\ & \left\{ \pi D_{\mathrm{c}} \sum_{i=0}^{n} q_{\mathrm{sui}} L_i + \frac{\pi}{4}(D_{\mathrm{c}}^2 - d_{\mathrm{c}}^2) q_{\mathrm{pu}} - B_{\mathrm{fw}}[(W_{\mathrm{con}} + W_{\mathrm{col}})x + W_{\mathrm{lh}} + W_{\mathrm{tool}}] \right. \end{aligned} \tag{1.16}$$

(6) Stability Criterion to Prevent Subsidence:

$$Q_{\mathrm{t}}^* \geq Q_{\mathrm{w}} \tag{1.17}$$

In the formula, Q_0 is the initial bearing capacity of the conductor, kN; Q_{t} is the real-time bearing capacity of the conductor, kN; Q_{t}^* is the initial bearing capacity of the conductor after a certain recovery time when it is lowered in place, kN; Q_{u} is the ultimate bearing capacity of the conductor, kN; Q_{w} is the total load after suspending the surface casing when the conductor is lowered in place, kN; R is the weight-on-bit coefficient, ranging from 0.8 to 1.0; B_{fw} is the buoyancy coefficient in seawater; x is the designed length of the conductor, m; W_{con} is the linear weight of the conductor, kN/m; W_{col} is the linear weight of the jetting drilling tool, kN/m; W_{lh} is the gravity of the low-pressure wellhead, kN; W_{tool} is the gravity of the running tool, kN; B_{fd} is the buoyancy coefficient in the drilling fluid; W_{scr} is the linear weight of the surface casing, kN/m; L_{sc} is the running depth of the surface casing, m; W_{hh} is the gravity of the high-pressure wellhead, kN; D_{c} is the outer diameter of the conductor, m; d_{c} is the inner diameter of the conductor, m; q_{sui} is the ultimate side resistance per unit area of the i soil layer around the conductor, kPa; q_{pu} is the ultimate end resistance per unit area, kPa; L_i is the thickness of the i-th soil layer, m; n is the number of soil layers; α_b is the bearing capacity growth coefficient; t is the recovery time, d.

1.2 Analysis of the Characteristic Parameters of the Mechanical Stability of Subsea Wellheads and Their Influencing Factors

According to the previous analysis, the main types of instability of subsea wellheads in deepwater drilling are wellhead inclination and wellhead subsidence. For the force analysis of wellhead inclination, refer to Fig. 1.1. The subsea wellhead is placed above the conductor. The lateral external force and bending moment that cause its inclination come from the flexible/spherical joint between the Blowout Preventer (BOP) and the riser, while the lower conductor plays a role in supporting the wellhead. Therefore, the characteristic parameters of the inclination of the subsea wellhead are the lateral displacement and offset angle at the wellhead. For the force analysis of wellhead subsidence, refer to Fig. 1.2. The subsea wellhead is placed above the conductor and suspends the surface casing at the same time. The main cause of the wellhead subsidence is that the bearing capacity of the conductor is insufficient to support the buoyant weight of the subsea wellhead and the surface casing. Therefore, the characteristic parameter of the subsidence of the subsea wellhead is the difference between the buoyant weight of the string composed of the subsea wellhead, conductor, and surface casing and the bearing capacity of the conductor.

In order to explore the influence rules of environmental loads, construction parameters and other factors on the characteristic parameters of the mechanical stability of subsea wellheads in deepwater drilling, so as to screen out sensitive and uncertain factors, the calculation model of the mechanical stability of subsea wellheads introduced in Sect. 1.1.1 is used to analyze the characteristic parameters of the inclination and subsidence of subsea wellheads and their influencing factors respectively.

1.2.1 Analysis of the Characteristic Parameters of the Lateral Mechanical Stability of Subsea Wellheads and Their Influencing Factors

A deepwater well in the South China Sea is taken as the research object to study the characteristic parameters of the lateral mechanical stability of subsea wellheads and their influencing factors. The basic parameters of this deepwater well are shown in Table 1.1 as follows.

Table 1.1 The basic parameters of deepwater well

Environmental loads	Values	Engineering parameters	Values	Material properties	Values
Platform average drift/(% water depth)	3	Water depth/m	1525	Elastic modulus of steel/GPa	210
Wind—driven current velocity on water surface/(m·s^{-1})	1.2	Drilling fluid density/(kg·m^{-3})	1180	Density of steel/(kg·m^{-3})	7850
Tidal current velocityon water surface/(m·s^{-1})	0.6	Length of riser/m	1525	Outer diameter of riser/mm	533.4
Drag coefficient	0.8	Tension ratio of riser	1.5	Wall thickness of riser/mm	25.4
Seawater density/(kg·m^{-3})	1030	Setting depth of conductor pipe/m	80	Linear weight of jetting drill string/(kN·m)	3.6
Buoyancy coefficient in seawater	0.85	Distance between wellhead and mud line/m	3.2	Outer diameter of conductor pipe/mm	914.4
Number of seabed soil layers	10	Hanging weight at wellhead/kN	3000	Wall thickness of conductor pipe/mm	25.4
Initial undrained shear strength/kPa	20	Setting depth of surface casing/m	685	Outer diameter of surface casing/mm	508
Increment of undrained shear strength at 10 m/kPa	15	Height of blowout preventer (BOP) stack/m	18	Wall thickness of surface casing/mm	12.7
Initial underwater unit weight of soil/(kg·m^{-3})	7.0	Weight of BOP stack/kN	2000	Linear weight of surface casing/(kN·m)	2.1
Underwater unit weight of soil at 10 m/(kg·m^{-3})	0.5	Equivalent diameter of BOP/m	1	Elastic modulus of cement sheath/GPa	18

1.2.1.1 Influence of Different Factors on the Characteristic Parameters of the Lateral Mechanical Stability of Subsea Wellheads

Use the calculation model of the lateral mechanical stability of subsea wellheads introduced in Sect. 1.1.1 to calculate the characteristic parameters of the lateral mechanical stability of the subsea wellhead of this deepwater well: the displacement and offset angle of the subsea wellhead. Then, respectively change the environmental loads and construction parameters such as water depth, sea current velocity at the water surface, tidal current velocity at the water surface, drag coefficient, inertial force coefficient, average platform drift, riser tension ratio, height of the subsea wellhead from the mud line, running depth of the conductor, outer diameter of the conductor, wall thickness of the conductor, running depth of the surface casing, etc., and recalculate the displacement and offset angle of the subsea wellhead.

In order to make the calculation results better reflect the influence of various factors on the characteristic parameters, the following regulations are made for the values of each factor x_i:

(1) Combining with engineering practice and actual needs, determine the value range $\left[x_{i,\max}, x_{i,\min}\right]$;
(2) Determine the number of values n of x_i according to the requirements of calculation accuracy, then:

$$x_{ij} = x_{i,\min} + \frac{x_{i,\max} - x_{i,\min}}{n}(j-1) \tag{1.18}$$

where x_{ij} is the j value of the factor x_i, and $j = 1, 2, 3, \cdots, n$.

(3) When the factor x_i has $m(m > 1)$ fixed sizes or values in engineering, if $m = 2$, let x_1 and x_2 be the boundary values of the value range of x_i, and use Eq. (1.18) to take values for x_i; if $m > 2$, then let x_1 and x_{m} be the boundary values of the value range of x_i, and use Eq. (1.18) to take values for x_i.

For this deepwater well, select $n = 5$ to calculate the characteristic parameters of the lateral mechanical stability of the subsea wellhead under different conditions, and the results are shown in Table 1.2 as follows.

As can be seen from Table 1.2, the running depth of the conductor within the value interval of [70, 90]m and the increment of the underwater unit weight of the seabed soil per 10 m have no influence on the displacement and offset angle of the subsea wellhead. The increment of the undrained shear strength of the seabed soil per 10 m has basically no influence on the displacement and offset angle of the subsea wellhead, while other factors all have an impact on the displacement and offset angle of the subsea wellhead.

In order to further explore the influence of the running depth of the conductor on the lateral mechanical stability of the subsea wellhead, the influence of the characteristic parameters when the running depth of the conductor is relatively shallow is calculated, as shown in Table 1.3. At the same time, the increment of the underwater unit weight of the seabed soil per 10 m has no influence on the wellhead displacement and offset angle, which indicates that the influence of the underwater unit weight of the soil on the mechanical stability of the wellhead is only limited to the interval of 0–10 m. The fact that the increment of the undrained shear strength per 10 m has little influence on the mechanical stability of the wellhead shows that the influence of the undrained shear strength of the soil on the mechanical stability of the wellhead is also limited to the relatively shallow interval of the seabed soil. Therefore, in order to further study the influence of the properties of the seabed soil on the lateral mechanical stability of the wellhead, the increments of the underwater unit weight and the undrained shear strength of the seabed soil are refined, and the influence

Table 1.2 The characteristic parameters of horizontal mechanical stability of subsea wellhead under different conditions

Water depth/m	Wellhead displacement/m	Deviation angle/(°)
1325	0.276	2.461
1425	0.309	2.727
1525	0.342	2.983
1625	0.375	3.244
1725	0.409	3.501
Wind—driven current velocity on water surface/(m·s^{-1})	Wellhead displacement/m	Deviation angle/(°)
0.8	0.189	1.744
1.0	0.257	2.322
1.2	0.342	2.983
1.4	0.447	3.773
1.6	0.571	4.683
Tidal current velocity on water surface/(m·s^{-1})	Wellhead displacement/m	Deviation angle/(°)
0.4	0.303	2.684
0.5	0.321	2.828
0.6	0.342	2.983
0.7	0.363	3.144
0.8	0.385	3.311
Drag coefficient	Wellhead displacement/m	Deviation angle/(°)
0.6	0.259	2.336
0.7	0.299	2.653
0.8	0.342	2.983
0.9	0.386	3.319
1.0	0.431	3.656
Lifting force of buoyancy blocks/kN	Wellhead displacement/m	Deviation angle/(°)
1750	0.360	3.129
2000	0.351	3.057
2250	0.342	2.983
2500	0.332	2.906
2750	0.324	2.834
Platform average drift/(% Water depth)	Wellhead displacement/m	Deviation angle/(°)
1	0.271	2.432
2	0.306	2.707
3	0.342	2.983
4	0.379	3.268
5	0.418	3.554

(continued)

Table 1.2 (continued)

Water depth/m	Wellhead displacement/m	Deviation angle/(°)
Tension ratio of riser	Wellhead displacement/m	Deviation angle/(°)
1.1	0.130	1.270
1.3	0.244	2.204
1.5	0.342	2.983
1.7	0.434	3.708
1.9	0.516	4.374
Setting depth of conductor pipe/m	Wellhead displacement/m	Deviation angle/(°)
70	0.342	2.983
75	0.342	2.983
80	0.342	2.983
85	0.342	2.983
90	0.342	2.983
Height from subsea wellhead to mud line/m	Wellhead displacement/m	Deviation angle/(°)
2.6	0.324	2.898
2.8	0.329	2.923
3.0	0.337	2.962
3.2	0.342	2.983
3.4	0.347	3.010
Outer diameter of conductor pipe/mm	Wellhead displacement/m	Deviation angle/(°)
762.0	0.637	5.661
800.1	0.533	4.731
838.2	0.455	4.019
876.3	0.393	3.451
914.4	0.342	2.983
Wall thickness of conductor pipe/mm	Wellhead displacement/m	Deviation angle/(°)
25.400	0.342	2.983
28.575	0.315	2.735
31.750	0.294	2.535
34.925	0.276	2.369
38.100	0.261	2.225
Density of drilling fluid/(kN·m^{-3})	Wellhead displacement/m	Deviation angle/(°)
1080	0.331	2.899
1180	0.342	2.983
1280	0.352	3.064
1380	0.361	3.141

(continued)

Table 1.2 (continued)

Water depth/m	Wellhead displacement/m	Deviation angle/(°)
1480	0.370	3.217
Outer diameter of riser/mm	Wellhead displacement/m	Deviation angle/(°)
406.4	0.217	1.982
457.2	0.274	2.374
508	0.324	2.770
558.8	0.375	3.167
609.6	0.417	3.571
Elastic modulus of cement sheath/GPa	Wellhead displacement/m	Deviation angle/(°)
8	0.370	3.252
13	0.356	3.112
18	0.342	2.983
23	0.330	2.870
28	0.318	2.761
Initial underwater unit weight/(kN·m^{-3})	Wellhead displacement/m	Deviation angle/(°)
4	0.354	3.037
5.5	0.348	3.010
7	0.342	2.984
8.5	0.336	2.957
10	0.330	2.930
Initial undrained shear strength/kPa	Wellhead displacement/m	Deviation angle/(°)
10	0.441	3.415
15	0.385	3.175
20	0.342	2.983
25	0.311	2.841
30	0.287	2.728
Increment of underwater unit weight at 10 m/(kN·m^{-3})	Wellhead displacement/m	Deviation angle/(°)
0.1	0.342	2.983
0.3	0.342	2.983
0.5	0.342	2.983
0.7	0.342	2.983
0.9	0.342	2.983
Increment of undrained shear strength at 10 m/kPa	Wellhead displacement/m	Deviation angle/(°)
5	0.345	3.005
10	0.342	2.991
15	0.342	2.983

(continued)

Table 1.2 (continued)

Water depth/m	Wellhead displacement/m	Deviation angle/(°)
20	0.341	2.977
25	0.339	2.966

Table 1.3 The characteristic parameters of horizontal mechanical stability of subsea wellhead with different conductor setting depth

Setting depth of conductor pipe/m	Wellhead displacement/m	Deviation angle/(°)
5	1.7	12.549
10	0.717	5.052
15	0.391	3.205
18	0.347	3.009
20	0.342	2.983
25	0.342	2.983
30	0.342	2.983

of these two factors on the mechanical stability of the wellhead is recalculated, as shown in Table 1.4.

As can be seen from Table 1.3, when the running depth of the conductor is less than 20 m, with the decrease of the running depth of the conductor, the displacement and offset angle of the subsea wellhead increase significantly. When the running depth of the conductor is greater than 20 m, the wellhead displacement and offset angle are no longer affected by the running depth of the conductor. In practical engineering applications, the running depth of the conductor in deepwater drilling is much greater than 20 m, so it can be considered that the running depth of the conductor has no influence on the lateral mechanical stability of the subsea wellhead.

As can be seen from Table 1.4, the refined increment of the underwater unit weight of the soil still has little influence on the displacement and offset angle of the subsea wellhead and can be basically ignored, indicating that the increment

Table 1.4 The characteristic parameters of horizontal mechanical stability of subsea wellhead with different soil bulk density increment and non-drainage shear strength increment

Increment of underwater unit weight of soil per meter/(kN·m^{-3})	Wellhead displacement/m	Deviation angle/(°)	Increment of undrained shear strength of soil per meter/kPa	Wellhead displacement/m	Deviation angle/(°)
0.01	0.343	2.993	0.5	0.334	2.955
0.03	0.342	2.992	1	0.325	2.920
0.05	0.342	2.991	1.5	0.315	2.883
0.07	0.342	2.990	2	0.309	2.855
0.09	0.342	2.989	2.5	0.304	2.830

of the underwater unit weight of the soil has basically no influence on the lateral mechanical stability of the subsea wellhead. The refined increment of the undrained shear strength of the soil has a greater influence on the displacement and offset angle of the subsea wellhead, indicating that the influence of the increment of the undrained shear strength of the seabed soil on the wellhead displacement and offset angle occurs in the shallow formation. After further calculation and verification, when the soil depth is greater than 30 m, the magnitude of the increment of the undrained shear strength no longer has an influence on the wellhead displacement and offset angle. Therefore, the influence of the increment of the undrained shear strength on the lateral mechanical stability of the subsea wellhead is limited to the shallow seabed formation interval of [0, 30]m.

1.2.1.2 Sensitivity Analysis of the Influence of Various Factors on the Lateral Mechanical Stability of Subsea Wellheads

In order to comprehensively study the sensitivity of different factors to the lateral mechanical stability of subsea wellheads, it is first necessary to carry out a dimensionless treatment of each influencing factor x_{ij}. The commonly used dimensionless methods are:

(1) "Centralization" dimensionless method

$$x_{ij}{\prime} = \frac{x_{ij} - \overline{x}_j}{S_j} \tag{1.19}$$

In the formula, $\overline{x}_j$ is the mean value of x_j, and S_j is the standard deviation of x_j.

This dimensionless method removes the influence of dimensions and magnitudes and is a commonly used dimensionless method. However, this method also removes the variability between variables and is not applicable to the situation where the variables are not uniformly distributed and the variability between the original variables needs to be reflected after dimensionless treatment.

(2) "Range normalization" dimensionless method

$$x_{ij}{\prime} = \frac{x_{ij} - x_{j,\min}}{x_{j,\max} - x_{j,\min}} \tag{1.20}$$

In the formula, $x_{j,\min}$ and $x_{j,\max}$ are the minimum and maximum values of x_j respectively.

(3) "Maximization" dimensionless method

$$x_{ij}{\prime} = \frac{x_{ij}}{x_{j,\min}}(x_{j,\min} > 0) \tag{1.21}$$

(4) "Minimization" dimensionless method

$$x_{ij}\prime = \frac{x_{ij}}{x_{j,\max}} \tag{1.22}$$

(5) "Averaging" dimensionless method

$$x_{ij}\prime = \frac{x_{ij}}{\bar{x}_j} \tag{1.23}$$

This dimensionless method can retain the variability among various factors and has a wide range of applications.

Among the above five dimensionless methods, (1) and (5) are more widely used, and the dimensionless treatment has little impact on the comprehensive evaluation results.

Since the values of x_{ij} are specified in Sect. 1.2.1.1 (see formula 1.18), and the values of each factor belong to the uniformly distributed type, either the dimensionless method (1) or (5) can be used for the dimensionless treatment. In order to make the values of x_{ij} be in the same interval after dimensionless treatment, this book uses the "centralization" dimensionless method to perform dimensionless treatment on the data in Tables 1.2 and 1.4, and the results are shown in Table 1.5.

In order to have a more intuitive understanding of the influence of each dimensionless factor on the lateral mechanical stability of the subsea wellhead, the data in Table 1.5 are plotted. The dimensionless factors are taken as the abscissa, and the displacement or offset angle of the subsea wellhead is taken as the ordinate, as shown in Figs. 1.3 and 1.4.

As can be seen from Figs. 1.3 and 1.4, there are significant differences in the influence of various dimensionless factors on the displacement and offset angle of the subsea wellhead. However, for a single dimensionless factor, its influence on the displacement and offset angle of the subsea wellhead is basically consistent. Most of the factors are linearly correlated, and only the influence of individual factors is nonlinear with a small degree of nonlinearity, which can be approximately regarded as a linear correlation. At the same time, the increment of the underwater unit weight per meter has basically no influence on the displacement and offset angle of the subsea wellhead, and the curve in the Figure approaches a horizontal straight line. While the six dimensionless factors, namely the outer diameter of the conductor, the wall thickness of the conductor, the lifting force of the buoyancy block, the elastic modulus of the cement sheath, the initial undrained shear strength, and the increment of the undrained shear strength per meter, are negatively correlated with the displacement and offset angle of the subsea wellhead.

In order to quantitatively analyze the influence of each factor on the lateral mechanical stability of the subsea wellhead, it is necessary to conduct a linear regression analysis on the influence laws of each factor on the wellhead displacement and offset angle in Figs. 1.3 and 1.4, and the least squares method can be used.

Table 1.5 The characteristic parameters of horizontal mechanical stability of subsea wellhead with non-dimensional factors

Water depth/m	Wellhead displacement/m	Deviation angle/(°)
−1.265	0.276	2.461
−0.632	0.309	2.727
0.000	0.342	2.983
0.632	0.375	3.244
1.265	0.409	3.501
Wind—driven current velocity on water surface/($m \cdot s^{-1}$)	Wellhead displacement/m	Deviation angle/(°)
−1.265	0.189	1.744
−0.632	0.257	2.322
0.000	0.342	2.983
0.632	0.447	3.773
1.265	0.571	4.683
Tidal current velocity on water surface/($m \cdot s^{-1}$)	Wellhead displacement/m	Deviation angle/(°)
−1.265	0.303	2.684
−0.632	0.321	2.828
0.000	0.342	2.983
0.632	0.363	3.144
1.265	0.385	3.311
Drag coefficient	Wellhead displacement/m	Deviation angle/(°)
−1.265	0.259	2.336
−0.632	0.299	2.653
0.000	0.342	2.983
0.632	0.386	3.319
1.265	0.431	3.656
Lifting force of buoyancy blocks/kN	Wellhead displacement/m	Deviation angle/(°)
−1.265	0.36	3.129
−0.632	0.351	3.057
0.000	0.342	2.983
0.632	0.332	2.906
1.265	0.324	2.834
Average platform drift/(% Water depth)	Wellhead displacement/m	Deviation angle/(°)
−1.265	0.271	2.432
−0.632	0.306	2.707
0.000	0.342	2.983
0.632	0.379	3.268
1.265	0.418	3.554

(continued)

Table 1.5 (continued)

Water depth/m	Wellhead displacement/m	Deviation angle/(°)
Tension ratio of riser	Wellhead displacement/m	Deviation angle/(°)
−1.265	0.13	1.27
−0.632	0.244	2.204
0.000	0.342	2.983
0.632	0.434	3.708
1.265	0.516	4.374
Height from wellhead to mud line/m	Wellhead displacement/m	Deviation angle/(°)
−1.265	0.324	2.898
−0.632	0.329	2.923
0.000	0.337	2.962
0.632	0.342	2.983
1.265	0.347	3.01
Outer diameter of conductor pipe/mm	Wellhead displacement/m	Deviation angle/(°)
−1.265	0.637	5.661
−0.632	0.533	4.731
0.000	0.455	4.019
0.632	0.393	3.451
1.265	0.342	2.983
Wall thickness of conductor pipe/mm	Wellhead displacement/m	Deviation angle/(°)
−1.265	0.342	2.983
−0.632	0.315	2.735
0.000	0.294	2.535
0.632	0.276	2.369
1.265	0.261	2.225
Density of drilling fluid/kg·m^{-3}	Wellhead displacement/m	Deviation angle/(°)
−1.265	0.331	2.899
−0.632	0.342	2.983
0.000	0.352	3.064
0.632	0.361	3.141
1.265	0.37	3.217
Outer diameter of riser/mm	Wellhead displacement/m	Deviation angle/(°)
−1.265	0.217	1.982
−0.632	0.274	2.374
0.000	0.324	2.770
0.632	0.375	3.167
1.265	0.417	3.571

(continued)

Table 1.5 (continued)

Water depth/m	Wellhead displacement/m	Deviation angle/(°)
Elastic modulus of cement sheath/GPa	Wellhead displacement/m	Deviation angle/(°)
−1.265	0.37	3.252
−0.632	0.356	3.112
0.000	0.342	2.983
0.632	0.33	2.87
1.265	0.318	2.761
Initial underwater unit weight/kN·m^{-3}	Wellhead displacement/m	Deviation angle/(°)
−1.265	0.354	3.037
−0.632	0.348	3.010
0.000	0.342	2.984
0.632	0.336	2.957
1.265	0.330	2.930
Initial undrained shear strength/Kpa	Wellhead displacement/m	Deviation angle/(°)
−1.265	0.441	3.415
−0.632	0.385	3.175
0.000	0.342	2.983
0.632	0.311	2.841
1.265	0.287	2.728
Increment of underwater unit weight per 1 m	Wellhead displacement/m	Deviation angle/(°)
−1.265	0.342	2.983
−0.632	0.342	2.983
0.000	0.342	2.983
0.632	0.342	2.983
1.265	0.342	2.983
Increment of undrained shear strength per 1 m	Wellhead displacement/m	Deviation angle/(°)
−1.265	0.345	3.005
−0.632	0.342	2.991
0.000	0.342	2.983
0.632	0.341	2.977
1.265	0.339	2.966

Taking the influence of the linearly regressed water depth H_i on the displacement S_{wi} of the subsea wellhead as an example, assume that the optimal linear fitting formula is $S_{\mathrm{wi}}\prime = aH_i + b$. According to the principle of the least squares method, it is necessary to satisfy:

$$S = \sum\left[S_{\mathrm{wi}} - (b + aH_i)^2\right] = \min \tag{1.24}$$

In the formula, S is the sum of the squares of the deviations between S_{wi} and $S_{\mathrm{wi}}\prime$. That is:

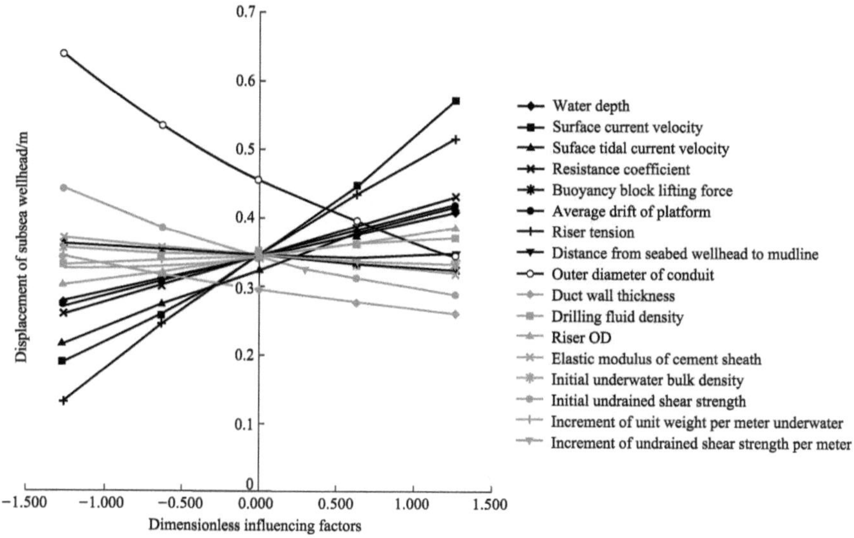

Fig. 1.3 The influence of non-dimensional factors on subsea wellhead displacement

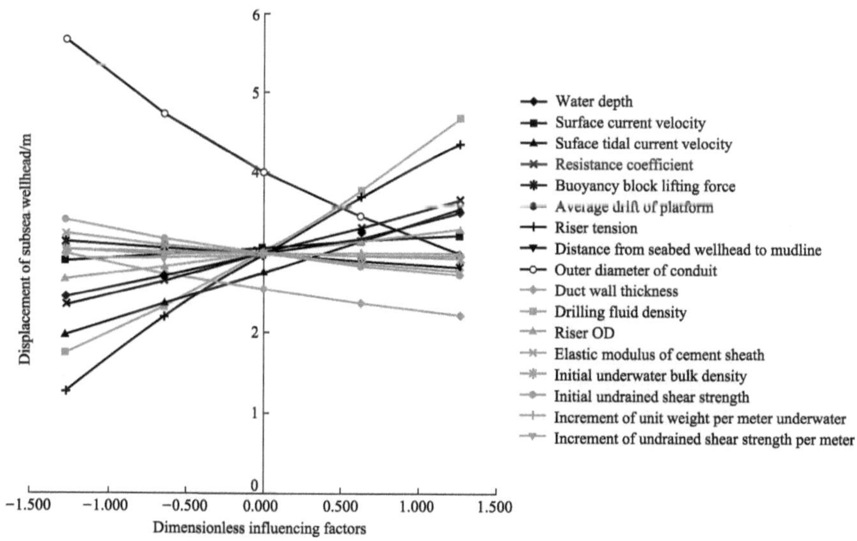

Fig. 1.4 The influence of non-dimensional factors on subsea wellhead deviation angle

$$\frac{\partial S}{\partial a} = -2\sum (S_{\mathrm{wi}} - b - aH_i) = 0 \tag{1.25}$$

$$\frac{\partial S}{\partial b} = -2\sum (S_{\mathrm{wi}} - b - aH_i)H_i = 0 \tag{1.26}$$

It can be transformed into:

$$R = \frac{\sum(H_i - \overline{H}_i)\sum(S_{\mathrm{wi}} - \overline{S}_{\mathrm{wi}})}{\sqrt{\sum(H_i - \overline{H}_i)^2}\sqrt{\sum(S_{\mathrm{wi}} - \overline{S}_{\mathrm{wi}})^2}} \tag{1.27}$$

$$\sum H_i S_{\mathrm{wi}} - b\sum H_i - a\sum H_i^2 = 0 \tag{1.28}$$

Solve to get:

$$a = \frac{\sum H_i \sum S_{\mathrm{wi}} - n\sum H_i S_{\mathrm{wi}}}{\left(\sum H_i\right)^2 - n\sum H_i^2} \tag{1.29}$$

$$b = \frac{\sum H_i S_{\mathrm{wi}} \sum H_i - \sum S_{\mathrm{wi}} \sum H_i^2}{\left(\sum H_i\right)^2 - n\sum H_i^2} \tag{1.30}$$

Thus, according to the original data regression, the linear regression equation between the water depth H_i and the displacement S_{wi} of the subsea wellhead is obtained:

$$S_{\mathrm{wi}}\prime = \frac{\sum H_i \sum S_{\mathrm{wi}} - n\sum H_i S_{\mathrm{wi}}}{\left(\sum H_i\right)^2 - n\sum H_i^2} H_i + \frac{\sum H_i S_{\mathrm{wi}} \sum H_i - \sum S_{\mathrm{wi}} \sum H_i^2}{\left(\sum H_i\right)^2 - n\sum H_i^2} \tag{1.31}$$

In order to characterize the degree of coincidence between the linear regression equation and the original data, the correlation coefficient R is introduced, and its value range is [−1,1].

$$R = \frac{\sum(H_i - \overline{H}_i)\sum(S_{\mathrm{wi}} - \overline{S}_{\mathrm{wi}})}{\sqrt{\sum(H_i - \overline{H}_i)^2}\sqrt{\sum(S_{\mathrm{wi}} - \overline{S}_{\mathrm{wi}})^2}} \tag{1.32}$$

R^2 is the coefficient of determination, which takes values within the range of 0 to 1, indicating the degree of correlation between $S_{\mathrm{wi}}\prime$ and S_{wi}. The closer it is to 1, the more correlated they are.

In Eqs. (1.31) and (1.32), H_i, a, and b can be replaced by x_i, a_i, and b_i respectively for the linear regression of other dimensionless factors.

Carry out a linear regression on the data in Table 1.5 according to Eq. (1.31), and calculate the corresponding coefficient of determination according to Eq. (1.32):

$$S_w \begin{cases} 0.0525x_1 + 0.3422(R^2 = 1) \\ 0.1508x_2 + 0.3612(R^2 = 0.986) \\ 0.0326x_3 + 0.3428(R^2 = 0.999) \\ 0.0681x_4 + 0.3434(R^2 = 1) \\ -0.0144x_5 + 0.3418(R^2 = 0.999) \\ 0.058x_6 + 0.3432(R^2 = 0.99) \\ 0.1521x_7 + 0.3332(R^2 = 0.996) \\ 0.0093x_8 + 0.3358(R^2 = 0.992) \\ -0.1154x_9 + 0.472(R^2 = 0.980) \\ -0.0318x_{10} + 0.2976(R^2 = 0.987) \\ 0.0153x_{11} + 0.3512(R^2 = 0.998) \\ 0.0829x_{12} + 0.3237(R^2 = 1) \\ -0.0206x_{13} + 0.3432(R^2 = 0.998) \\ -0.0095x_{14} + |0.342(R^2 = 1) \\ -0.0604x_{15} + 0.3532(R^2 = 0.972) \\ 10^{-7}x_{16} + 0.342(R^2 = 0.999) \\ -0.0022x_{17} + 0.342(R^2 = 0.980) \end{cases} \quad \theta_w \begin{cases} 0.4106x_1 + 2.9832(R^2 = 1) \\ 1.1588x_2 + 3.101(R^2 = 0.992) \\ 0.2482x_3 + 2.99(R^2 = 0.999) \\ 0.5227x_4 + 2.9894(R^2 = 1) \\ -0.1172x_5 + 2.9818(R^2 = 1) \\ 0.4435x_6 + 2.9888(R^2 = 1) \\ 1.2194x_7 + 2.9078(R^2 = 0.996) \\ 0.0449x_8 + 2.9552(R^2 = 0.992) \\ -1.0492x_9 + 4.169(R^2 = 0.982) \\ -0.2976x_{10} + 2.5694(R^2 = 0.988) \\ 0.1255x_{11} + 3.0608(R^2 = 1) \\ 0.6249x_{12} + 2.7709(R^2 = 1) \\ -0.1935x_{13} + 2.9956(R^2 = 0.997) \\ -0.0422x_{14} + 2.9836(R^2 = 1) \\ -0.2701x_{15} + 3.0284(R^2 = 0.978) \\ -10^{-16}x_{16} + 2.983(R^2 = 0.999) \\ -0.0145x_{17} + 2.9844(R^2 = 0.981) \end{cases} \tag{1.33}$$

In the formula, S_w is the displacement of the subsea wellhead, m; θ_w is the offset angle of the subsea wellhead, degrees; $x_i (i = 1, 2, 3, \cdots\cdots, 15, 16, 17)$ are respectively the dimensionless water depth, sea current velocity at the water surface, tidal current velocity at the water surface, drag coefficient, lifting force of the buoyancy block, average platform drift, riser tension, distance from the subsea wellhead to the mud line, outer diameter of the conductor, wall thickness of the conductor, density of the drilling fluid, outer diameter of the riser, elastic modulus of the cement sheath, initial underwater unit weight, initial undrained shear strength, increment of the underwater unit weight per meter, increment of the undrained shear strength per meter. The value interval of x_i is all $[-1.265, 1.265]$.

Next, according to the linear regression equation in Eq. (1.33), $y_i = a_i x_i + b_i$ (y_i represents S_w or θ_w), the sensitivity of each factor to the displacement and offset angle of the subsea wellhead is analyzed.

As can be seen from Figs. 1.3 and 1.4, each factor x_i has both positive and negative correlations with y_i, and the absolute value of the slope of each straight line directly reflects the sensitivity of x_i to y_i within the interval $[-1.265, 1.265]$. That is, the larger the absolute value of the slope of the straight line corresponding to x_i, the greater the change of y_i caused by changing the same value of the dimensionless factor (the value of the dimensionless factor changes within the interval $[-1.265, 1.265]$, and the value of the original data changes by the corresponding percentage within the corresponding interval). Therefore, $|a_i|$ can be used as an index to evaluate the sensitivity of x_i to y_i, and $|a_i|$ is defined as the sensitivity factor of x_i to y_i.

At the same time, the integral Y_i of each straight line within the interval $[x_{i\min}, x_{i\max}]$ (see Eq. 1.34) reflects the overall level of the influence of x_i on y_i, that is, the magnitude of the overall influence of this x_i on y_i within the interval $[x_{i\min}, x_{i\max}]$. The larger Y_i is, the more likely the factor x_i is to cause the displacement and angular offset of the subsea wellhead. Therefore, Y_i can be used as an index to evaluate the relative magnitude of the influence of x_i on y_i.

$$Y_i = \int_{x_{i\min}}^{x_{i\max}} (ax_i + b)\mathrm{d}x_i \tag{1.34}$$

Since the value intervals of x_i are all symmetric about the y-axis, when comparing the Y_i values corresponding to each factor x_i, Y_i can be replaced by the intercept b_i, and the principle is as follows:

Suppose the value interval of $[-c, c]$. Solving Eq. (1.34) gives:

$$Y_i = \int_{-c}^{c} (ax_i + b)\mathrm{d}x_i = \left(\frac{a}{2}x_i^2 + bx_i\right)\Big|_{-c}^{c} = 2bc \tag{1.35}$$

Then:

$$\frac{Y_i}{Y_j} = \frac{b_i c}{b_j c} = \frac{b_i}{b_j} \tag{1.36}$$

It can be seen that the magnitude relationship among Y_i can be replaced by the intercept b_i. According to the definition and significance of Y_i, b_i is defined as the influence baseline of x_i on y_i. Its physical meaning is the corresponding y_i. when the factor x_i takes the middle value within the value interval $[x_{i\min}, x_{i\max}]$, and its magnitude reflects the average level of the influence of x_i on y_i.

In order to better illustrate the meanings of $|a_i|$ and b_i, the influence of the outer diameter of the conductor and the wall thickness of the conductor of the example deepwater well on the displacement of the subsea wellhead will be described.

In Fig. 1.5, the sensitivity factor $|a_i| = 0.1154$ of the outer diameter of the conductor is obviously greater than the sensitivity factor $|a_i| = 0.0318$ of the wall thickness of the conductor, which indicates that the change of S_w caused by the corresponding proportional change of the outer diameter of the conductor within the value interval [762, 914.4] is greater than the change of S_w caused by the corresponding proportional change of the wall thickness of the conductor within the value interval [25.4, 38.1]. At the same time, the influence baseline $b_i = 0.472$ of the outer diameter of the conductor is also greater than $b_i = 0.2976$ of the wall thickness of the conductor, indicating that within their respective value intervals, the outer diameter of the conductor contributes more to the displacement of the subsea wellhead than the wall thickness of the conductor and has a higher average influence.

Therefore, $|a_i|$ and b_i can be used as the sensitivity factor and influence baseline of x_i on y_i in the linear regression equation $y_i = a_i x_i + b_i$ (y_i represents S_w or θ_w) respectively.

Calculate and rank the sensitivity index $|a_i|$ and influence baseline b_i of x_i on y_i in Eq. (1.33), as shown in Table 1.6.

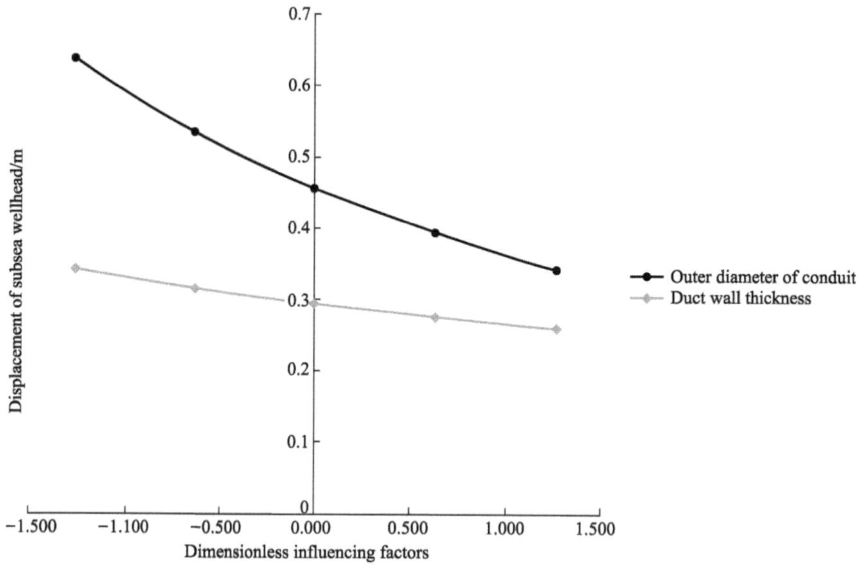

Fig. 1.5 The influence of conductor diameter and wall thickness on subsea wellhead displacement

According to the calculation results in Table 1.6, the following rankings can be obtained:

(1) The sensitivities ($|a_i|$) of various factors to the displacement S_w of the subsea wellhead, from largest to smallest, are: riser tension, sea current velocity at the water surface, outer diameter of the conductor, outer diameter of the riser, drag coefficient, initial undrained shear strength, average platform drift, water depth, tidal current velocity at the water surface, wall thickness of the conductor, elastic modulus of the cement sheath, density of the drilling fluid, lifting force of the buoyancy block, initial underwater unit weight, distance from the subsea wellhead to the mud line, increment of the undrained shear strength per meter, increment of the underwater unit weight per meter.

(2) The degrees of influence (b_i) of various factors on the displacement S_w of the subsea wellhead, from largest to smallest, are: outer diameter of the conductor, sea current velocity at the water surface, initial undrained shear strength, density of the drilling fluid, drag coefficient, average platform drift, elastic modulus of the cement sheath, tidal current velocity at the water surface, water depth, initial underwater unit weight, increment of the underwater unit weight per meter, increment of the undrained shear strength per meter, lifting force of the buoyancy block, distance from the subsea wellhead to the mud line, riser tension, outer diameter of the riser, wall thickness of the conductor.

(3) The sensitivities ($|a_i|$) of various factors to the offset angle θ_w of the subsea wellhead, from largest to smallest, are: riser tension, sea current velocity at the water surface, outer diameter of the conductor, outer diameter of the riser, drag

Table 1.6 The sensitivity factor $|a_i|$ and influence baseline bi as well as their ranking

Dimensionless factors	Displacement of subsea wellhead S_w				Deviation angle of subsea wellhead θ_w			
	$\|a_i\|$	Sort by $\|a_i\|$	$\|b_i\|$	Sort by $\|b_i\|$	$\|a_i\|$	Sort by $\|a_i\|$	$\|b_i\|$	Sort by $\|b_i\|$
Water depth	0.0525	8	0.3422	9	0.4106	7	2.9832	11
Wind—driven Current velocity on water surface	0.1508	2	0.3612	2	1.1588	2	3.101	2
Tidal current Velocity on water surface	0.0326	9	0.3428	8	0.2482	10	2.99	6
Drag coefficient	0.0681	5	0.3434	5	0.5227	5	2.9894	7
Lifting force of buoyancy blocks	0.0144	13	0.3418	13	0.1172	13	2.9818	13
Average platform drift	0.058	7	0.3432	6	0.4435	6	2.9888	8
Tension of riser	0.1521	1	0.3332	15	1.2194	1	2.9078	15
Distance from subsea wellhead to mud line	0.0093	15	0.3358	14	0.0449	14	2.9552	14
Outer diameter of conductor pipe	0.1154	3	0.472	1	1.0492	3	4.169	1
Wall thickness of conductor pipe	0.0318	10	0.2976	17	0.2976	8	2.5694	17
Density of drilling fluid	0.0153	12	0.3512	4	0.1255	12	3.0608	3
Outer diameter of riser	0.0829	4	0.3237	16	0.6249	4	2.7709	16
Elastic modulus of cement sheath	0.0206	11	0.3432	7	0.1935	11	2.9956	5
Initial underwater unit weight	0.0095	14	0.342	10	0.0422	15	2.9836	10
Initial undrained shear strength	0.0604	6	0.3532	3	0.2701	9	3.0284	4

(continued)

Table 1.6 (continued)

Dimensionless factors	Displacement of subsea wellhead S_w				Deviation angle of subsea wellhead θ_w			
	$\|a_i\|$	Sort by $\|a_i\|$	$\|b_i\|$	Sort by $\|b_i\|$	$\|a_i\|$	Sort by $\|a_i\|$	$\|b_i\|$	Sort by $\|b_i\|$
Increment of underwater unit weight per meter	10^{-7}	17	0.342	11	10^{-16}	17	2.983	12
Increment of undrained shear strength per meter	0.0022	16	0.342	12	0.0145	16	2.9844	9

coefficient, average platform drift, water depth, wall thickness of the conductor, initial undrained shear strength, tidal current velocity at the water surface, elastic modulus of the cement sheath, density of the drilling fluid, lifting force of the buoyancy block, distance from the subsea wellhead to the mud line, initial under water unit weight, increment of the undrained shear strength per meter, increment of the underwater unit weight per meter.

(4) The degrees of influence (b_i) of various factors on the offset angle θ_w of the subsea wellhead, from largest to smallest, are: outer diameter of the conductor, sea current velocity at the water surface, density of the drilling fluid, initial undrained shear strength, elastic modulus of the cement sheath, tidal current velocity at the water surface, drag coefficient, average platform drift, increment of the undrained shear strength per meter, initial underwater unit weight, water depth, increment of the underwater unit weight per meter, lifting force of the buoyancy block, distance from the subsea wellhead to the mud line, riser tension, outer diameter of the riser, wall thickness of the conductor.

It is worth noting that the sensitivity factors and influence baselines of the above rankings are all calculated based on the example deepwater well in this paper. The rankings are affected by the value intervals $[x_{i\min}, x_{i\max}]$ selected for each factor. In order to ensure the applicability of the ranking results, it is necessary to fully consider the actual situation on site and strictly select values according to the value selection method specified in Sect. 1.2.1.1.

According to the relative order of magnitude and magnitude of $|a_i|$ and b_i, the factors that have a greater influence on the displacement and offset angle of the subsea wellhead of the example deepwater well are screened out as follows:

(1) The six factors most sensitive to the displacement of the subsea wellhead are: riser tension > sea current velocity at the water surface > outer diameter of the conductor > outer diameter of the riser > drag coefficient > initial undrained shear strength.

(2) The six factors that have the greatest influence on the displacement of the subsea wellhead are: outer diameter of the conductor > sea current velocity at the water

surface > initial undrained shear strength > density of the drilling fluid > drag coefficient > average platform drift.

(3) The six factors most sensitive to the offset angle of the subsea wellhead are: riser tension > sea current velocity at the water surface > outer diameter of the conductor > outer diameter of the riser > drag coefficient > average platform drift.

(4) The six factors that have the greatest influence on the offset angle of the subsea wellhead are: outer diameter of the conductor > sea current velocity at the water surface > density of the drilling fluid > initial undrained shear strength > elastic modulus of the cement sheath > tidal current velocity at the water surface.

The larger the influence baseline b_i is, the greater the average influence level of x_i on y_i is, and the more unfavorable it is to the lateral mechanical stability of the subsea wellhead. The larger the sensitivity factor $|a_i|$ is, the more favorable it is to reduce y_i by adjusting x_i, which is the primary factor to be considered for improving the lateral mechanical stability of the subsea wellhead. Therefore, for the example deepwater well, the following conclusions can be drawn by synthesizing Table 1.6.

(1) Displacement of the subsea wellhead

Riser tension: Its corresponding $|a_i|$ value is the largest, and it is a factor that is relatively sensitive to the lateral mechanical stability of the subsea wellhead. The positive value of a_i indicates that the smaller the riser tension is, the more favorable it is to the lateral mechanical stability of the wellhead. Therefore, on the basis of ensuring the control of the riser deformation and improving the force condition, the top tension of the riser should be reduced as much as possible.

Outer diameter of the conductor: Its corresponding b_i value is the largest, and $|a_i|$ ranks third. It is a factor that has a relatively large and sensitive influence on the lateral mechanical stability of the subsea wellhead. The negative value of a_i indicates that the larger the outer diameter of the conductor is, the more favorable it is to the lateral mechanical stability of the wellhead. Therefore, a conductor with a large outer diameter should be selected as much as possible.

Sea current velocity at the water surface: Its corresponding b_i and $|a_i|$ both rank second. It is a factor that has a relatively large and sensitive influence on the lateral mechanical stability of the subsea wellhead. The positive value of a_i indicates that the smaller its value is, the more favorable it is to the lateral mechanical stability of the wellhead. However, this factor is an environmental load and is irresistible. Therefore, the sea current velocity at the water surface should be closely monitored, and other favorable parameters should be adjusted in a timely manner.

Initial undrained shear strength: Its corresponding b_i value ranks third, and $|a_i|$ ranks sixth. It is a factor that has a relatively large and sensitive influence on the lateral mechanical stability of the subsea wellhead. The negative value of a_i indicates that the larger its value is, the more favorable it is to the lateral mechanical stability of the wellhead. Therefore, the area with a large initial undrained shear strength of the shallow seabed soil should be selected for operation as much as possible.

Density of the drilling fluid: Its corresponding b_i value ranks fourth, and $|a_i|$ ranks twelfth. It is a factor that has a relatively large but not sensitive influence on the lateral mechanical stability of the subsea wellhead. This type of factor inevitably has an impact on the lateral mechanical stability of the subsea wellhead, and it is necessary to adjust other parameters with strong sensitivity (larger $|a_i|$) to improve the lateral mechanical stability of the subsea wellhead.

Drag coefficient: Its corresponding b_i and $|a_i|$ both rank fifth. It is a factor that has a relatively large and sensitive influence on the lateral mechanical stability of the subsea wellhead. This factor is an empirical coefficient, and a suitable value needs to be selected according to the region and engineering experience.

Average platform drift: Its corresponding b_i value ranks sixth, and $|a_i|$ ranks seventh. It is a factor that has a relatively large and sensitive influence on the lateral mechanical stability of the subsea wellhead. The positive value of a_i indicates that the smaller its value is, the more favorable it is to the lateral mechanical stability of the wellhead. Therefore, the average platform drift should be controlled within a relatively small range.

(2) Offset angle of the subsea wellhead

The influence laws of the riser tension, outer diameter of the conductor, sea current velocity at the water surface, density of the drilling fluid, and initial undrained shear strength on the lateral mechanical stability of the subsea wellhead are the same as those on the wellhead displacement.

Elastic modulus of the cement sheath: Its corresponding b_i value ranks fifth, and $|a_i|$ ranks eleventh. It is a factor that has a relatively large but not very sensitive influence on the lateral mechanical stability of the subsea wellhead. The positive value of a_i indicates that the smaller its value is, the more favorable it is to the lateral mechanical stability of the wellhead. Since its sensitivity is not high, it is not recommended to be the primary adjustment parameter for improving the lateral mechanical stability of the subsea wellhead.

Tidal current velocity at the water surface: Its corresponding b_i value ranks sixth, and $|a_i|$ ranks tenth. It is a factor that has a relatively large but not very sensitive influence on the lateral mechanical stability of the subsea wellhead. The positive value of a_i indicates that the smaller its value is, the more favorable it is to the lateral mechanical stability of the wellhead. However, as this factor is an environmental load, it is necessary to conduct real-time monitoring to adjust other parameters in a timely manner.

1.2.2 Analysis of the Characteristic Parameters of the Longitudinal Mechanical Stability of Subsea Wellheads and Their Influencing Factors

As shown in Fig. 1.2, the subsea wellhead is placed above the conductor and suspends the surface casing at the same time. The main cause of the wellhead subsidence is that the bearing capacity of the conductor is insufficient to support the buoyant weight of the subsea wellhead and the surface casing. Therefore, the characteristic parameter of the subsidence of the subsea wellhead is the difference between the buoyant weight of the string composed of the subsea wellhead, conductor, and surface casing and the bearing capacity of the conductor, that is, $\Delta Q = Q_t^* - Q_w$.

Since as long as ΔQ is greater than 0, the longitudinal mechanical stability of the subsea wellhead can be ensured, it is not convenient to be used as an index for quantitatively evaluating the longitudinal mechanical stability of the subsea wellhead. And the factors determining the magnitude of ΔQ are mainly the factors determining the real-time bearing capacity Q_t^* of the conductor after a certain recovery time when it is lowered in place. According to Eq. (1.16), the factors determining the magnitude of Q_t^* mainly include: the buoyancy coefficients of seawater and drilling fluid, the gravity of the combined string, the outer diameter of the conductor, the running depth of the conductor, the side resistance and end resistance of the conductor in the shallow seabed soil, the bearing capacity growth coefficient, the recovery time, etc.

Among many factors, the running depth of the conductor plays a decisive role and is also a parameter that needs special attention in the design and construction of deepwater drilling. At the same time, the running depth of the conductor that can ensure $\Delta Q \geq 0$ also incorporates the influence of other factors such as the properties of the seabed soil, the bearing capacity growth coefficient, and the recovery time. Therefore, the running depth $H_{\min}$ of the conductor when the running depth $H_{\min}$ of the conductor when $\Delta Q = \varepsilon$ (ε is the safety margin) can be used as the characteristic parameter of the longitudinal mechanical stability of the subsea wellhead.

For the convenience of calculation, the upper and lower limit values of the safety margin are set:

$$\varepsilon_d < Q_t^* - Q_w < \varepsilon_u \tag{1.37}$$

Then the minimum running depth $H_{\min}$ of the conductor that meets the bearing capacity requirement can be jointly calculated by Eqs. (1.13), (1.16), and (1.37).

1.2.2.1 Influence of Different Factors on the Characteristic Parameters of the Longitudinal Mechanical Stability of Subsea Wellheads

Take the conductor running operation of a deepwater well in the South China Sea as the research object to study the characteristic parameters of the longitudinal mechanical stability of subsea wellheads and their influencing factors. The basic parameters of the conductor jetting running operation of this deepwater well and the resistance parameters of the seabed soil to the conductor are shown in Tables 1.7 and 1.8 respectively.

Jointly calculate the running depth H_{min} of the conductor that meets the bearing capacity using Eqs. (1.13), (1.16), and (1.37). Select the lower limit ε_d of the safety margin as 50 kN and the upper limit ε_u as 100 kN. Then, through iterative calculation, the running depth of the conductor $H_{min} = 75$ m m is obtained.

According to the force analysis diagram of the conductor when the surface casing is suspended after the conductor is run in place in Fig. 1.2, and by analyzing the parameters in Table 1.7, it can be known that the buoyant weights of the jet drilling tool assembly, high-pressure wellhead, low-pressure wellhead, running tool, and surface casing are all the longitudinal loads of the conductor, and these parameters are engineering parameters designed in advance, which should be regarded as fixed values here. While the outer diameter of the conductor, the wall thickness of the conductor, the bearing capacity growth coefficient of the seabed soil, and the recovery

Table 1.7 The basic parameters of setting conductor

Outer diameter of conductor pipe/mm	914.4	Length of suspended surface casing/m	850
Wall thickness of conductor pipe/mm	25.4	Linear weight of surface casing/(kN·m)	2.1
Linear weight of conductor pipe/(kN·m)	7.8	Mass of running tool/kg	3.1×10^3
Linear weight of jetting drill string combination/(kN·m)	3.6	Buoyancy coefficient in water	0.85
Mass of low—pressure wellhead/kg	13.1×10^3	Bearing capacity growth coefficient of seabed soil	0.1
Mass of high—pressure wellhead/kg	4.1×10^3	Soil recovery time/d	1

Table 1.8 Resistance parameters of subsea soil to conductor

Depth below mud line/m	Unit side resistance/kPa	Unit end resistance/kPa
10	8	100
20	20	150
30	45	250
40	60	400
50	80	650
60	95	800
70	105	950
80	125	1000
90	130	1050
100	145	1100
110	160	1100
120	165	1300

time are the construction parameters of the conductor during jetting running, and these parameters may all affect the bearing capacity of the conductor after it is run in place, that is, the longitudinal stability of the wellhead.

Therefore, according to the value regulation of Eq. (1.18), take $n = 5$, and calculate the running depth of the conductor when the outer diameter of the conductor, the wall thickness of the conductor, the bearing capacity growth coefficient, and the recovery time are different respectively. The specific calculation results are shown in Table 1.9.

To more clearly reflect the influence law of each factor on the penetration depth of the conductor, the data in Table 1.9 are plotted as shown in Fig. 1.6.

As can be seen from the above figures, the wall thickness of the conductor has no influence on the running depth of the conductor. The outer diameter of the conductor has a relatively small influence on the running depth of the conductor. As the outer diameter of the conductor increases, the running depth of the conductor decreases linearly. The bearing capacity growth coefficient has a significant influence on the running depth of the conductor. Especially when the value of the bearing capacity growth coefficient is relatively small, as the bearing capacity growth coefficient decreases, the running depth of the conductor increases sharply. The recovery time also has a great influence on the running depth of the conductor. When the recovery time is less than 1.5d, the running depth of the conductor increases sharply as the recovery time decreases, and when the recovery time is greater than 1.5d, the influence is relatively small.

Table 1.9 The setting depths of conductor with different parameters

Outer diameter of conductor pipe/mm	Setting depth of conductor pipe/m	Wall thickness of conductor pipe/mm	Setting depth of conductor pipe/m
762.0	83	25.400	75
800.1	81	28.575	75
838.2	79	31.750	75
876.3	77	34.925	75
914.4	75	38.100	75
Bearing capacity growth coefficient	Setting depth of conductor pipe/m	Recovery time/d	Setting depth of conductor pipe/m
0.05	95	0.5	92
0.10	75	1	75
0.15	60	1.5	68
0.20	53	2	65
0.25	47	2.5	63

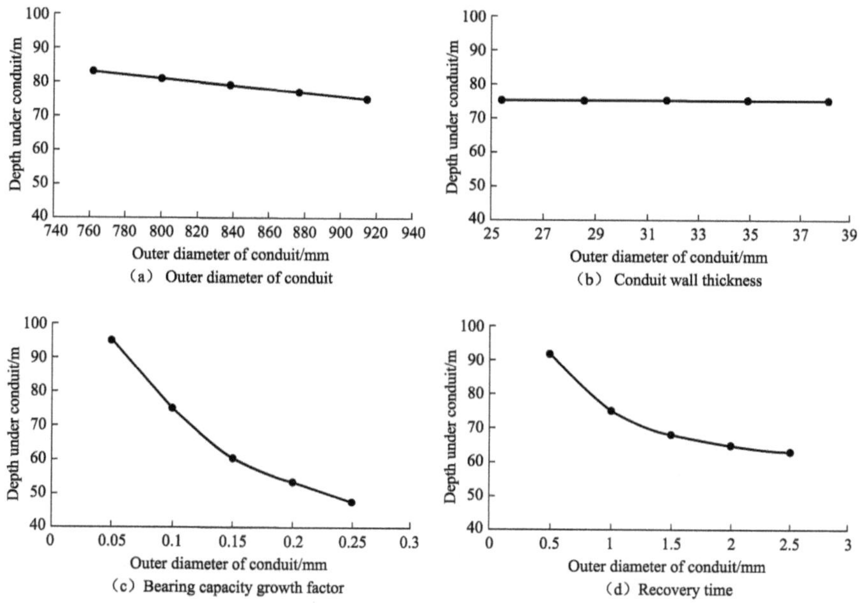

Fig. 1.6 The influence of different parameters on conductor setting depth

1.2.2.2 Sensitivity Analysis of the Influence of Various Factors on the Longitudinal Mechanical Stability of Subsea Wellheads

In order to comprehensively study the sensitivity of different factors to the longitudinal mechanical stability of subsea wellheads, the method in Sect. 1.2.1.2 is adopted to carry out a dimensionless treatment of the data in Table 1.9, and the results are shown in Table 1.10.

Plot the data in Table 1.10, with the dimensionless factors as the abscissa and the running depth of the conductor as the ordinate, as shown in Fig. 1.7.

Table 1.10 The characteristic parameters of vertical mechanical stability of subsea wellhead with non-dimensional factors

Outer diameter of conductor pipe/mm	Setting depth of conductor pipe/m	Bearing capacity growth coefficient	Setting depth of conductor pipe/m	Recovery time	Setting depth of conductor pipe/m
−1.265	83	−1.265	95	−1.265	92
−0.632	81	−0.632	75	−0.632	75
0.000	79	0.000	60	0.000	68
0.632	77	0.632	53	0.632	65
1.265	75	1.265	47	1.265	63

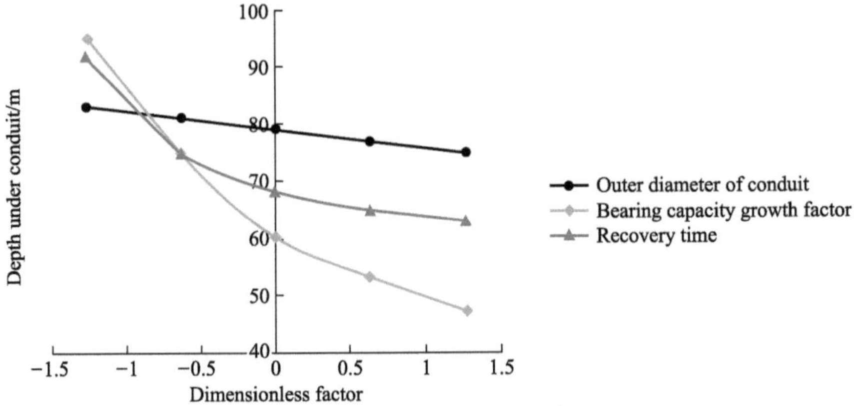

Fig. 1.7 The influence of different non-dimensional factors on conductor setting depth

As can be seen from Fig. 1.7, the bearing capacity growth coefficient has the greatest influence on the running depth of the conductor. Within the value interval [0.05, 0.25], the variation range of the running depth of the conductor reaches 42 m. Followed by the recovery time, when the recovery time is short, its influence on the running depth of the conductor is consistent with that of the bearing capacity growth coefficient. As the recovery time prolongs, its influence on the running depth of the conductor gradually decreases. Within the value interval [0.5, 2.5]d, the variation range of the running depth of the conductor is 29 m. The outer diameter of the conductor has a relatively small influence on the running depth of the conductor, which is basically linearly correlated. Within the value interval [762, 914.4]mm, the variation range of the running depth of the conductor is 8 m.

Therefore, the bearing capacity growth coefficient and the recovery time need to be fully considered in the design of the running depth of the conductor. It is necessary to determine the value of the bearing capacity growth coefficient as accurately as possible and appropriately extend the recovery time (not less than 1d). At the same time, the outer diameter of the conductor is also a sensitive factor affecting the lateral mechanical stability of the subsea wellhead. Selecting a conductor with a larger outer diameter is beneficial to improving both the lateral and longitudinal mechanical stability of the subsea wellhead.

1.3 Discrimination Method of the Instability Risk of Subsea Wellheads in Deepwater Drilling

Previously, the characteristic parameters of the mechanical stability of subsea wellheads and the sensitivity of their influencing factors were analyzed, and the factors that have a relatively large and sensitive influence on the mechanical stability of subsea wellheads were identified. We can start from these factors, find out the factors that are uncertain and relatively sensitive for research, and establish a discrimination method for the instability risk of subsea wellheads considering uncertain factors.

1.3.1 Discrimination Method of the Lateral Instability Risk of Subsea Wellheads

1.3.1.1 Analysis of Parameter Uncertainty

As can be seen from the analysis in Sect. 1.2.1.2, for the example deepwater well, the six factors most sensitive to the displacement of the subsea wellhead are: riser tension > sea current velocity at the water surface > outer diameter of the conductor > outer diameter of the riser > drag coefficient > initial undrained shear strength; the six factors most sensitive to the offset angle of the subsea wellhead are: riser tension > sea current velocity at the water surface > outer diameter of the conductor > outer diameter of the riser > drag coefficient > average platform drift.

Among many sensitive factors, the initial undrained shear strength is an environmental load, and relatively accurate results can be obtained through experiments and measurements; the outer diameter of the conductor, the outer diameter of the riser, and the riser tension are construction parameters, and their values can also be precisely controlled and determined; the sea current velocity at the water surface and the average platform drift are parameters that change with time. It is necessary to calculate the maximum values of the sea current velocity at the water surface and the average platform drift that the subsea wellhead can withstand through fitting formulas respectively, so as to reasonably control other construction parameters. The drag coefficient is an empirical coefficient in the calculation model of the lateral mechanical stability of the subsea wellhead in Sect. 1.1.1. The "Classification and Construction Specification for Mobile Offshore Units" stipulates that the value range of the drag coefficient is 0.6–1.2. According to the calculation of the example deepwater well, within this value range of the drag coefficient, the displacement variation range of the subsea wellhead is [0.259, 0.5158]m, and the offset angle variation range is [2.336°, 4.3118°]. It can be seen that when the drag coefficient is selected as 0.6 and 1.2 respectively, the calculated wellhead displacements differ by nearly 0.3 m, and the offset angles differ by nearly 2 degrees, indicating that its value has a great influence on the calculation of the characteristic parameters of the lateral mechanical stability of the subsea wellhead. It is necessary to incorporate the uncertainty of this coefficient into the risk discrimination of the wellhead instability.

The determination of the drag coefficient C_D is affected by many factors, such as the Reynolds number of the ocean current, the roughness of the riser wall, the structure of the riser system, and its attached organisms, etc. It is an empirical coefficient related to the marine environment and engineering structure. It is difficult to obtain an accurate fixed value for it, and it changes with different marine environments and working conditions. Therefore, this coefficient can be defined as a parameter containing probability information that follows a certain distribution form, and the wellhead characteristic parameters at different probabilities in this distribution form are calculated respectively, so as to incorporate the risk probability information caused by C_D into the discrimination of the instability risk of the lateral mechanical stability of the subsea wellhead, and avoid the instability risk caused by the inaccurate value of the drag coefficient to the greatest extent.

The forms of numerical distribution mainly include uniform distribution, triangular distribution, normal distribution, etc. According to the characteristics of the drag coefficient, this paper selects the normal distribution as its distribution form, mainly for the following two reasons:

(1) Under specific marine environmental and engineering structural conditions, the true value of the drag coefficient should tend to a specific value.
(2) The marine environment is constantly changing, and the roughness of the riser and the condition of its attached organisms are also constantly changing over time. During the entire deep—water drilling process, its value will fluctuate within a certain interval.

Then the probability density function and cumulative probability distribution function of the normal distribution of the drag coefficient C_D are respectively:

$$p(C_D) = \frac{1}{\sqrt{2\pi}\sigma} e^{-\frac{(C_D-\mu)^2}{2\sigma^2}} \tag{1.38}$$

$$F(C_D) = \frac{1}{\sqrt{2\pi}\sigma} \int_{-\infty}^{C_D} e^{-\frac{(t-\mu)^2}{2\sigma^2}} dt \tag{1.39}$$

In the formulas, μ is the mathematical expectation, representing the average value of C_D; σ is the standard deviation, representing the discreteness of the distribution of C_D. The smaller σ, the more concentrated C_D is distributed on both sides of μ.

According to the meaning of the drag coefficient C_D, μ can be defined as the central value of C_D, and its value needs to be comprehensively determined according to the marine environment and engineering structure conditions; σ can be defined as the fluctuation coefficient of C_D, and its value is comprehensively determined according to the degree of change in the marine environment and the requirements of risk management and control.

Taking the central value $\mu = 0.8$ as an example, the normal distribution forms of C_D with different fluctuation coefficients σ are calculated, as shown in Fig. 1.8.

It can be seen that different fluctuation coefficients can well reflect the degree of concentration of the drag coefficient relative to the central value. The smaller the fluctuation coefficient is, the more concentrated C_D is on both sides of μ, indicating that the smaller the value error of C_D is.

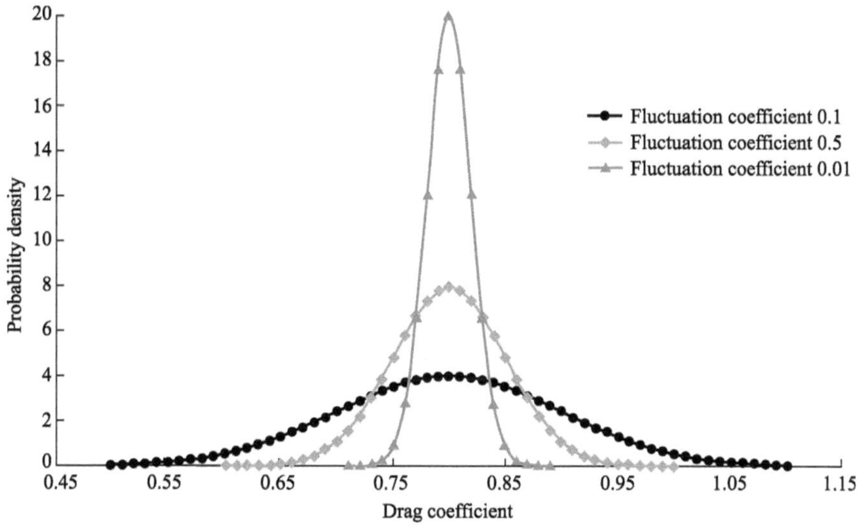

Fig. 1.8 Normal distributions with different fluctuation coefficient

According to Eq. (1.4) and the difference Eq. (1.9), the drag coefficient C_D has a linear correlation with both S_w and θ_w, that is, S_w and θ_w follow the distribution form of C_D and are normally distributed.

1.3.1.2 Discrimination Method of Wellhead Instability Risk

Based on the previous analysis of parameter uncertainty, the discrimination of the lateral instability risk of the subsea wellhead can be carried out according to the following steps:

(1) According to the requirements of risk management and control, determine the safety upper limits S_{wm} and θ_{wm} of the displacement and offset angle of the subsea wellhead. At the same time, according to the marine environment, engineering structure and risk management and control requirements of the target deepwater well, determine the central value μ and the fluctuation coefficient σ of the drag coefficient $C_D(\mu, \sigma)$ respectively, and its distribution form is shown in Fig. 1.8.
(2) Using the calculation results in Sect. 2.1.1, when the drag coefficient C_D takes the central value μ, fit the relationships between the sea current velocity at the water surface μ_w, the average platform drift S_O and the displacement and offset angle of the subsea wellhead respectively, and obtain the fitting formulas of each variable $S_w(\mu_w)$, $\theta_w(\mu_w)$, $S_w(S_O)$ and $\theta_w(S_O)$.
(3) Let $S_w(\mu_w) = S_{wm}$, $\theta_w(\mu_w) = \theta_{wm}$, $S_w(S_O) = S_{wm}$, $\theta_w(S_O) = \theta_w(\mu_w)$, and solve for μ_{w1}, μ_{w2}, S_{O1}, S_{O2} that meet the conditions respectively. Then the maximum sea current velocity $\mu_{wm} = \min\{\mu_{w1}, \mu_{w2}\}$, and the maximum average platform

drift $S_{0m} = \min\{S_{01}, S_{02}\}$ that ensure the lateral mechanical stability of the subsea wellhead can be obtained.

(4) According to the distribution form of C_D, change the value of C_D and repeat steps (2) and (3) to determine the distribution forms of μ_{wm} and S_{0m} respectively. Their probability density and cumulative probability distributions are shown in Figs. 1.9 and 1.10 respectively. Then, when the sea current velocity at the water surface and the average platform drift are μ_w and S_w respectively, the probability of the lateral instability risk of the subsea wellhead is:

$$P_h = \max\{J_u, J_S\} \tag{1.40}$$

In the formula, J_u and J_S are the cumulative probabilities corresponding to μ_w and S_w respectively, which represent the probabilities that the maximum sea current velocity or the maximum average platform drift is less than μ_w or S_w respectively.

(5) At the same time, for the convenience of on-site real-time monitoring, the risk threshold J_A (the safety upper limit of the wellhead instability risk, the larger the value, the lower the reliability) can be determined according to the actual situation. The corresponding sea current velocity at the water surface and the average platform drift are μ_{wA} and S_{wA} respectively. Then, when the monitored $\mu_w > \mu_{wA}$ or $S_w > S_{wA}$, it is considered that there is a risk of instability in the lateral mechanical stability of the subsea wellhead, and corresponding measures should be taken immediately to reduce the instability risk.

It should be noted that the safety upper limits S_{wm} and θ_{wm} of the displacement and offset angle of the subsea wellhead mentioned in step (1) of the discrimination method for the lateral instability risk of the subsea wellhead are the maximum allowable

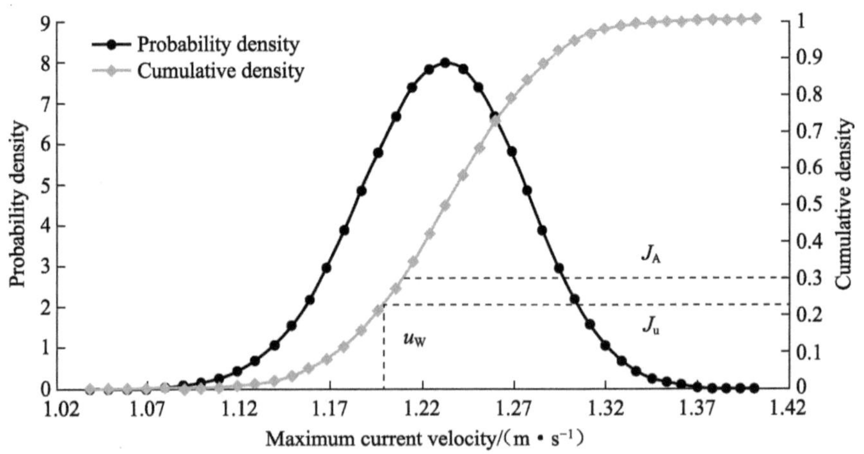

Fig. 1.9 The probability density and cumulative probability distribution of maximum current velocity

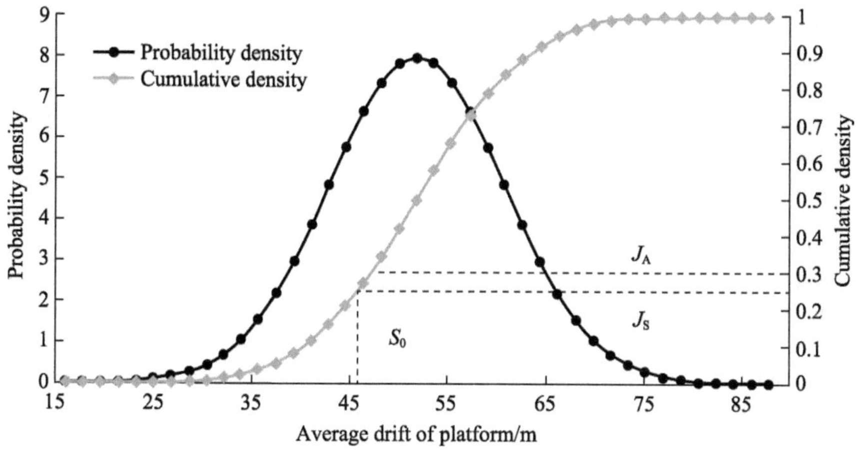

Fig. 1.10 The probability density and cumulative probability distribution of maximum platform average drift

values of the displacement and offset angle of the subsea wellhead to ensure the stability of the subsea wellhead and the safety of the riser. These values need to be determined by comprehensively considering various factors, including: whether the subsequent casing can be smoothly run in through the wellhead, the degree of eccentric wear of the drill pipe, the condition of the vortex-induced vibration of the riser, the distance of the wellhead above the mud line, etc.

1.3.2 *Discrimination Method of the Longitudinal Instability Risk of Subsea Wellheads*

1.3.2.1 Analysis of Parameter Uncertainty

(1) Ultimate Running Depth of the Conductor

According to the analysis in Sect. 1.1, the subsidence of the subsea wellhead is mainly due to the fact that the designed running depth of the conductor is too small, resulting in the bearing capacity of the conductor after it is run in place being insufficient to support the weight of the wellhead device and the subsequent casing, which usually occurs before the cementing of the surface casing. It should be noted that if the designed running depth of the conductor is too large in order to meet the bearing capacity requirement after the conductor is run in place, the conductor may not be able to be jetted in place normally. In on-site construction, if the situation of being unable to continue jetting and drilling occurs at a certain depth, measures such as moving the pipe string up and down, increasing the displacement, and adding bentonite drilling fluid are generally adopted to continue drilling. Although this makes the conductor string run in deeper, it will affect the sag of the conductor string, and there is a risk of wellhead instability when the conductor, after being run in place, bears the loads such as the high-pressure wellhead and the surface casing. Therefore, on the basis of studying the minimum running depth of the conductor, it is necessary to study the ultimate running depth of the conductor.

During the process of jetting the conductor into place, the most important control parameter in construction is the weight on bit, which is used to ensure that the conductor can be smoothly run in place and keep the pipe string vertical at the same time. The control of the weight on bit should ensure that the neutral point of the pipe string is below the mud line during the running process of the conductor, and the drilling is carried out by using the self-gravity of the pipe string entering the mud. At the same time, in order to avoid the bending of the pipe string due to excessive compression and to maximize the bearing capacity of the conductor after it is in place, it is required that the weight on bit when the conductor finally reaches the position should not be lower than 80% of the maximum weight on bit. In order to prevent the bending of the conductor string due to excessive compression during the construction process, the drilling can only be carried out by relying on the self-weight of the pipe string entering the mud. Therefore, there is a certain limit to the running depth of the conductor. If it exceeds this depth, the conductor cannot be run in place by relying on the self-gravity of the pipe string entering the mud.

Since the jetting during the running process of the conductor will disturb the shallow soil, the measured resistances at the end and on the side of the pipe determined by the soil test are not applicable to the calculation of the bearing capacity of the conductor just after it is run in place. According to the numerical simulation of the conductor jetting drilling in relevant literatures, the influence of the hydraulic jetting on the stress field of the surrounding soil mainly focuses near and below the nozzle. During the jetting drilling process, the soil near the nozzle undergoes deformation and damage due to the impact, and the resistances on the side and at the end of the pipe within a certain range at the lower end of the conductor can be ignored. Thus, the bearing capacity of the conductor when it is run in to the ultimate running depth is:

$$Q_{\mathrm{Lim}} = \pi D_{\mathrm{c}} \left(\sum_{i=0}^{x} q_{\mathrm{sui}} L_i - \sum_{j=k}^{y} q_{\mathrm{suj}} L_j \right) \tag{1.41}$$

In the formula, q_{sui} is the ultimate side resistance per unit area of the j soil layer affected by the jetting, kPa; y is the number of soil layers affected by the jetting; L_j is the length of the area where the soil outside the conductor is greatly affected by the jetting, m. Its value is related to the displacement, the weight on bit when the conductor reaches the position, the type of the conductor jetting drilling tool, and the jetting drilling speed, etc. It is caused by the fact that the soil disturbance caused by the jetting effect has not had time to recover, and it can be obtained from the results of numerical simulation.

According to the principle of mechanical equilibrium, when the conductor is run in to the ultimate position, it should satisfy:

$$
\begin{aligned}
Q_0 &= RB_{\text{fw}}[(W_{\text{con}} + W_{\text{col}})H_{\text{Lim}} + W_{\text{lh}} + W_{\text{tool}}] \\
&= \pi D_{\text{c}}\left(\sum_{i=0}^{x} q_{\text{sui}}L_i - \sum_{i=k}^{y} q_{\text{suj}}L_j\right) = Q_{\text{Lim}}
\end{aligned}
\tag{1.42}
$$

In the formula, H_{Lim} is the ultimate running depth of the conductor, m.

Equation (1.42) requires iterative calculation of H_{Lim}, and the minimum designed running depth $H_{\min}$ of the conductor can be selected as the initial value of H_{Lim}. According to Eq. (1.42), the expression for the ultimate running depth of the conductor can be obtained as:

$$
H_{\text{Lim}} = \frac{\frac{\pi D_{\text{c}}\left(\sum_{i=0}^{x} q_{\text{sui}}L_i - \sum_{i=k}^{y} q_{\text{suj}}L_j\right)}{RB_{\text{fw}}} - W_{\text{lh}} - W_{\text{tool}}}{W_{\text{con}} + W_{\text{col}}}
\tag{1.43}
$$

In Eq. (1.43), R is the weight on bit coefficient. When designing the running depth of the conductor, it takes a value between 0.8 and 1.0, and at this time, it is a fixed value. However, during the actual jetting drilling process, the weight on bit when the conductor reaches the position is affected by many factors, and the actual value fluctuates within a certain reasonable range. Therefore, when calculating the ultimate running depth of the conductor, R can no longer take the single value used in the design of the running depth, but fluctuates around this single value. Based on this characteristic, the distribution form of R can be set as $N(R, \sigma^2)$, where R is the weight on bit coefficient when designing the running depth of the conductor, and σ is determined according to the fluctuation range of the designed weight on bit.

(2) Bearing Capacity Growth Coefficient

As can be seen from the analysis in Sect. 1.2.2.2, the bearing capacity growth coefficient, the recovery time, and the outer diameter of the conductor are all sensitive factors affecting the longitudinal mechanical stability of the subsea wellhead, among which the bearing capacity growth coefficient and the recovery time have the most significant influence. The recovery time and the outer diameter of the conductor are construction parameters, which are known design parameters for the target deepwater well and are accurate fixed values. However, the bearing capacity growth coefficient is a regional empirical coefficient, and there is a large uncertainty in its value. A slight change in it will cause a large variation range of the designed running depth of the conductor. Therefore, it is necessary to make a reasonable prediction of the bearing capacity growth coefficient.

The bearing capacity growth coefficient has the characteristics of spatial limitation, continuity, and anisotropy in a region at the same time, and meets the characteristics of regional variables. Therefore, the geostatistics theory can be applied to make a regional prediction of it.

First, divide the region into grids, and use square grids, as shown in Fig. 1.11. Among them, σ_x and σ_y are reasonably valued according to the size of the region and the number of data points, and try to make each grid contain at least one data point. σ_z is the thickness of the grid. Since the bearing capacity growth coefficient has no change in the vertical direction, $\sigma_z = 0$ can be set, that is, a planar grid division is carried out.

Suppose the spatial expression of the bearing capacity growth coefficient in this region is $\alpha_i(x_i, y_i)$, where i is the number of known data points. Then the value of the bearing capacity growth coefficient of each grid is:

$$\beta_i(x_i, y_i) = \frac{1}{m}\sum_{i=1}^{m} \alpha_i(x_i, y_i) \tag{1.44}$$

Secondly, the variogram is obtained. The research object in this book is the parameters of a planar region, so only the method of obtaining the planar variogram is introduced.

Let A_i represent the planar coordinate (x_i, y_i), d represent the distance between the centers of the grids where two data points are located, then $\beta(A_i)$ and $\beta(A_i + d)$ respectively represent the values of the regional parameter at the positions A_i and $A_i + d$, where $i = 1, 2, \cdots, N(d)$, and $N(d)$ represents the total number of pairs of grids with a separation distance of d on the grid plane. According to the definition of the variogram, the formula for calculating the variogram is:

$$\gamma^{\#}(d) = \frac{1}{2N(d)}\sum_{i=1}^{N(d)} [\beta(A_i) - \beta(A_i + d)]^2 \tag{1.45}$$

Use Eq. (1.45) to calculate the variogram values $\gamma^{\#}(d)$ under different interval d conditions, and fit or apply them to the commonly used theoretical models of the variogram. The commonly used theoretical models of the variogram will not be elaborated here.

According to the Kriging method, the predicted value with probability information of the bearing capacity growth coefficient at the target point can be obtained:

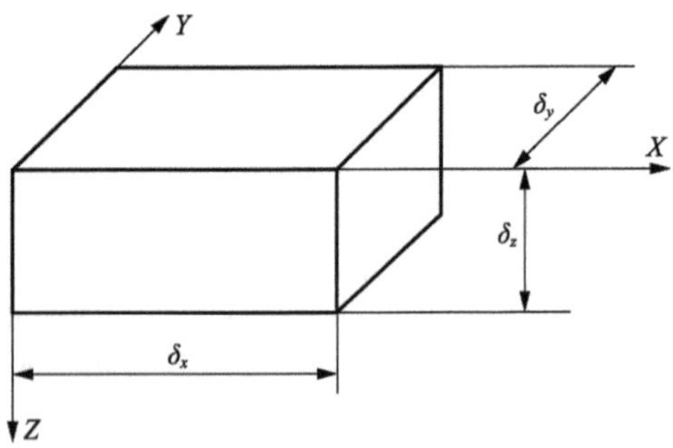

Fig. 1.11 Schematic diagram of regional grid

$$\beta(A^{\#}) = \sum_{i=1}^{n} \lambda_i \beta(A_i) \tag{1.46}$$

Among them, λ_i can be obtained by solving the Kriging equations:

$$\begin{cases} \sum_{j=1}^{n} \lambda_j \overline{C}[A_i, A_i] - \mu = \overline{C}[A_i, A^{\#}] \\ \sum_{i=1}^{n} \lambda_j = 1 \end{cases} \tag{1.47}$$

Among them, $\overline{C}[A_i, A_i] = \frac{1}{n}\sum_{i=1}^{n}\sum_{j=1}^{n} C(A_i - A_j)$, $\overline{C}[A_i, A^{\#}] = \frac{1}{n}\sum_{i=1}^{n} C(A_i - A^{\#})$.

In the formula, $C(x)$ is the covariance function of the variable x.

Calculate the variance of the predicted value of the bearing capacity growth coefficient at the target point obtained above, that is:

$$\sigma^2 = -\sum_{i=1}^{n} \lambda_i \overline{C}[A_i, A^{\#}] + \mu \tag{1.48}$$

In this way, the normal distribution of the bearing capacity growth coefficient at the target point is obtained, and its distribution form is $N(\beta(A^{\#}), \sigma^2)$.

It should be noted that the calculation accuracy of this method depends on the abundance of adjacent well data, such as the number, spacing and distribution of known well locations.

1.3.2.2 Discrimination Method of Wellhead Instability Risk

By using the calculation models of the minimum running depth and the ultimate running depth of the deepwater drilling conductor established previously, and combining with the uncertainty analysis of the bearing capacity growth coefficient and the weight on bit coefficient in the models, the uncertainty analysis and probabilistic treatment can be carried out for the boundaries of the running depth interval of the conductor (between the minimum running depth and the ultimate running depth).

First, through the regional prediction of the bearing capacity growth coefficient, the upper boundary of the running depth interval of the deepwater conductor can be realized as a probabilistic distribution. Suppose the probabilistic distribution form of the bearing capacity growth coefficient at the target well is $N(\beta(A^{\#}), \sigma^2)$, as shown in Fig. 1.12. According to the model, the probabilistic distribution form of the minimum running depth of the conductor can be obtained, as shown in Fig. 1.13.

In Fig. 1.13, the probability distribution of the minimum running depth of the conductor is no longer in a symmetric form. This is because the influence of the bearing capacity growth coefficient on the minimum running depth of the conductor

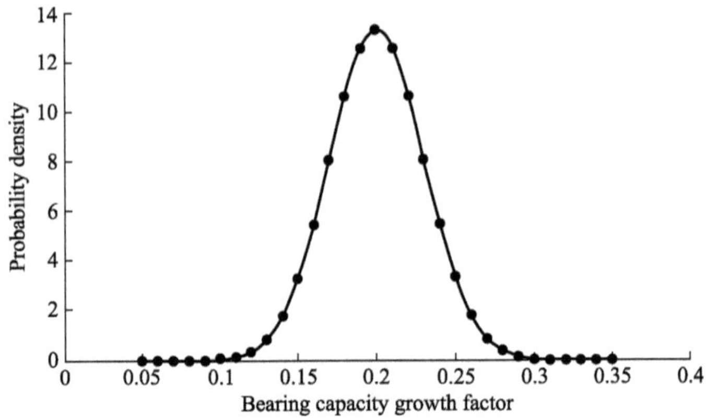

Fig. 1.12 Probability distribution of growth factor of target well

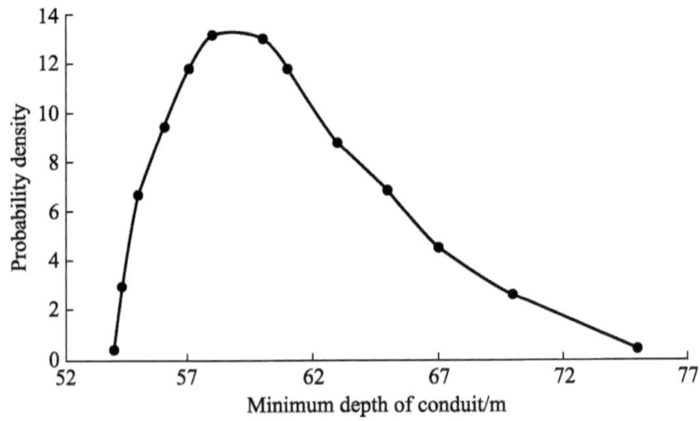

Fig. 1.13 Probability distribution of the minimum depth of conductor of target well

is in an approximately exponential form (see Fig. 1.6c). The smaller the bearing capacity growth coefficient is, the higher the change rate of the minimum running depth of the conductor is.

Secondly, for the lower boundary of the running depth interval of the conductor H_{Lim} assume that the distribution form of the weight on bit coefficient R in Eq. (1.43) is $N\left(R, \sigma^2\right)$, where R is the design value of the weight on bit coefficient when designing the running depth of the conductor, and σ is determined according to the fluctuation range of the weight on bit when the conductor is run in place during actual drilling. According to Eq. (1.43), the probabilistic distribution of the ultimate running depth of the conductor can be obtained, as shown in Fig. 1.14.

Thus, by comprehensively considering the probabilistic distributions of the minimum running depth and the ultimate running depth of the conductor, the risk

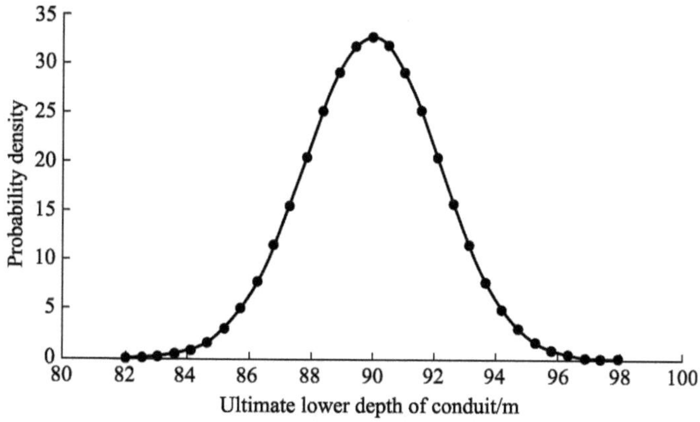

Fig. 1.14 Probability distribution of ultimate depth of conductor of target well

probability corresponding to the designed running depth of the conductor of the target deepwater well can be determined, as shown in Fig. 1.15a.

In order to facilitate the calculation of the safe designed running depth of the conductor under a certain confidence level condition or the risk assessment of the existing designed running depth of the conductor, the probability density in Fig. 1.15a is transformed into the cumulative probability in Fig. 1.15b. Given the confidence $J = J_2 - J_1$, as shown in Fig. 1.15b, then under the condition of the confidence level J, the risk assessment method for the given running depth of the conductor (longitudinal instability of the subsea wellhead) is as follows:

(1) If the designed running depth of the conductor is in the *BC* region, it is an absolutely safe region.
(2) If the designed running depth H_1 of the conductor is in the *AB* region, there is a risk that the wellhead will subside due to the insufficient bearing capacity of the conductor, that is, the risk of instability of the longitudinal mechanical stability. Specifically, the risk $P_v = F_1 - J_1$.
(3) If the designed running depth H_2 of the conductor is in the *CD* region, there is a risk that the conductor cannot be normally jetted and run in place. Specifically, the risk $P_{fail} = F_2 - J_1$.

1.4 Example Calculation

Using the discrimination method of the instability risk of subsea wellheads in deepwater drilling established in Sect. 1.3, the lateral instability risk and longitudinal instability risk of the subsea wellheads of the example deepwater wells in Sects. 2.1 and 2.2 are analyzed respectively.

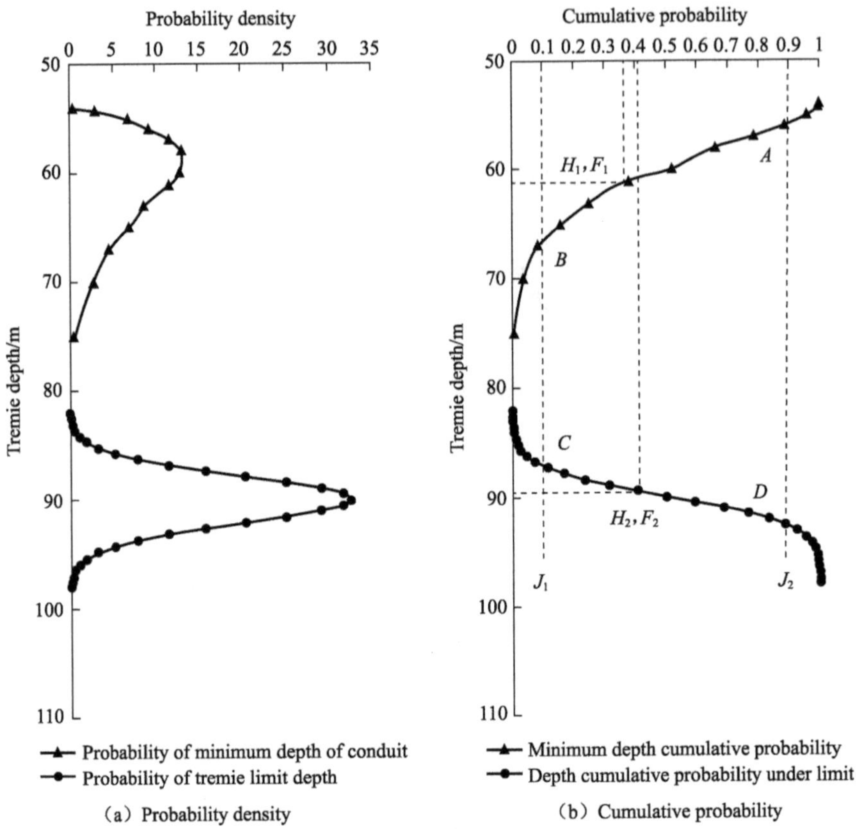

Fig. 1.15 Probability density and cumulative probability distribution of the minimum and ultimate depth of conductor for target well

1.4.1 Example Calculation of the Lateral Instability Risk of Subsea Wellheads

(1) Suppose that according to the requirements of risk management and control, the safety upper limit of the displacement of the subsea wellhead is determined as $S_{\text{wm}} = 0.37$ m, and the safety upper limit of the offset angle is determined as $\theta_{\text{wm}} = 3.1°$. At the same time, according to the marine environment, engineering structure and risk management and control requirements of the target deepwater well, the central value of the drag coefficient is determined as $\mu = 0.8$ and the fluctuation coefficient is determined as $\sigma = 0.05$, which satisfies the normal distribution of $C_{\text{D}}(0.8, 0.05)$.

(2) Using the calculation results in Sect. 2.1.1, when the drag coefficient C_{D} takes the central value of 0.8, fit the relationships between the sea current velocity at the water surface μ_{w}, the average platform drift S_{O} and the displacement

and offset angle of the subsea wellhead respectively, and the following fitting formulas are obtained:

$$S_w(u_w) = 0.2357u_w^2 - 0.0887u_w + 0.1094, \left(R^2 = 1\right)$$

$$S_w(S_O) = 0.0367S_O + 0.2331, (R^2 = 0.9996)$$

$$\theta_w(u_w) = 1.4161u_w^2 + 0.2659u_w + 0.6295, \left(R^2 = 1\right)$$

$$\theta_w(S_O) = 0.2805S_O + 2.1473, \left(R^2 = 0.9999\right)$$

(3) Let $S_w(u_w) = S_w(S_O) = 0.37\,\text{m}$, $\theta_w(u_w) = \theta_w(S_O) = 3.1°$, and solve for $u_{wl} = 1.26\,\text{m/s}$, $S_{O1} = 3.73(\%\text{water depth}) = 56.88\,\text{m}$, $u_{w2} = 1.23\,\text{m/s}$, $S_{O2} = 3.40(\%\text{water depth}) = 51.85\,\text{m}$ that meet the conditions respectively. Then the maximum sea current velocity $u_{wm} = \min\{1.26, 1.23\} = 1.23\,\text{m/s}$ that ensures the lateral mechanical stability of the subsea wellhead, and the maximum average platform drift $S_{Om} = \min\{56.88, 51.85\} = 51.85\,\text{m}$.

(4) According to the distribution form of C_D, change the value of C_D and repeat steps (2) and (3) to determine the distribution forms of u_w and S_{Om} respectively. Their probability density and cumulative probability distributions are shown in Figs. 1.16 and 1.10 respectively. Then, when the sea current velocity at the water surface and the average platform drift are $u_w = 1.2\,\text{m/s}$ and $S_w = 3\% = 45.75\,\text{m}$ respectively, the probability of the lateral instability risk of the subsea wellhead is: $P_h = \max\{J_u, J_S\} = \max\{0.22, 0.24\} = 0.24 = 24\%$.

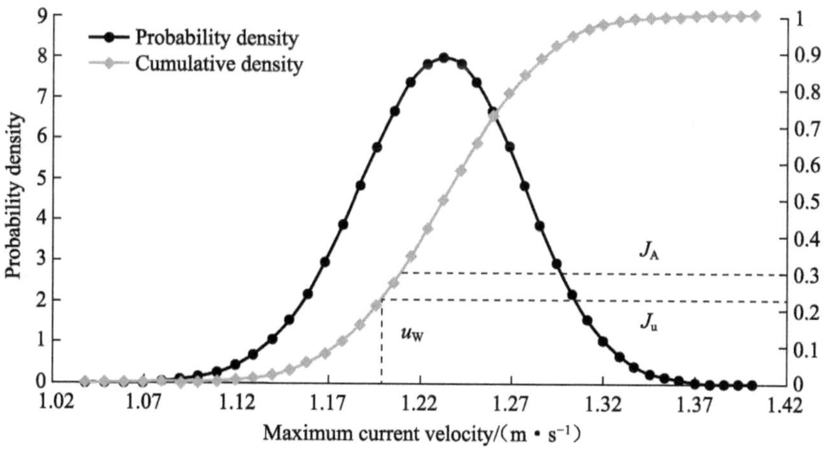

Fig. 1.16 The probability density and cumulative probability distribution of maximum current velocity of target well

(5) At the same time, for the convenience of on-site real-time monitoring, the risk threshold $J_A = 15\%$ (that is, the risk is controlled within 15%) is determined. The corresponding sea current velocity at the water surface and the average platform drift are $u_{wA} = 1.188\,\text{m/s}$ and $S_{wA} = 43\,\text{m}$ respectively. Then, when the monitored $u_{wA} > 1.188\,\text{m/s}$ or $S_{wA} > 43\,\text{m}$, it is considered that there is a risk of instability in the lateral mechanical stability of the subsea wellhead, and corresponding measures should be taken immediately to reduce the instability risk.

Thus, based on the selected central value of C_D ($\mu = 0.8$), the fluctuation coefficient $\sigma = 0.8$, the safety upper limit of the displacement of the subsea wellhead $S_{wm} = 0.37\,\text{m}$ and the safety upper limit of the offset angle $\theta_{wm} = 3.1°$, it is determined that there is a 24% probability of lateral instability of the subsea wellhead of the target deepwater well. At the same time, during the construction process, it should be ensured that $u_w < 1.188\,\text{m/s}$ and $S_w < 43\,\text{m}$.

In order to improve the lateral mechanical stability of the subsea wellhead of the example deepwater well, according to the analysis of the influence of various factors on the characteristic parameters of the lateral mechanical stability of the subsea wellhead in Sect. 2.1, the following measures can be adopted:

(1) Riser tension: On the basis of ensuring the control of the deformation of the riser and improving its stress state, the top tension of the riser should be reduced as much as possible. At the same time, this parameter is an important parameter for adjusting the lateral mechanical stability of the subsea wellhead, and it needs to be adjusted in a timely manner according to actual needs during the construction process.
(2) Outer diameter of the conductor: Try to select a conductor with a large outer diameter.
(3) Initial undrained shear strength: Try to select the area with a large initial undrained shear strength of the shallow seabed soil for operation.
(4) Average platform drift: During the construction process, the average platform drift should be controlled within a small range, and the drift of the platform needs to be monitored in real time to ensure that the average drift does not exceed $S_0 = 43\,\text{m}$.
(5) Sea current velocity at the water surface: During the construction process, the sea current velocity at the water surface should be closely monitored. Once it is found that $u_w > 1.188\,\text{m/s}$, it is necessary to timely and appropriately reduce the top tension of the riser, and try to control the platform drift as much as possible, etc.

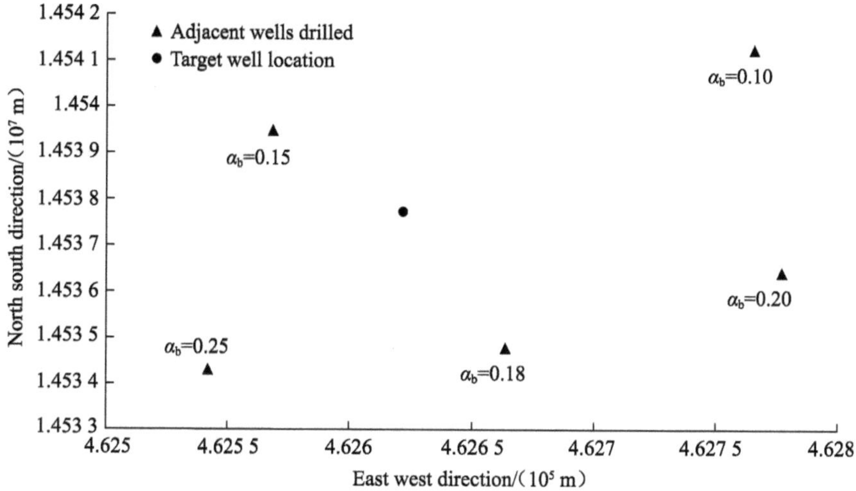

Fig. 1.17 Well location map

1.4.2 Example Calculation of the Longitudinal Instability Risk of Subsea Wellheads

The distribution of the drilled well locations and the target well location in the block of the example deepwater well in Sect. 1.2.2.1 is shown in Fig. 1.17. It is considered that the bearing capacity growth coefficient used for each drilled well is a safe and reliable value. The method in Sect. 1.3.2.2 is used to obtain the bearing capacity growth coefficient with probability information at the target well. This coefficient follows a normal distribution form with a mathematical expectation of 0.175 and a standard deviation of 0.03. During the jetting drilling process, the weight on bit coefficient R takes a value of 0.9, and the standard deviation takes a value of 0.02.

The methods in Sects. 1.2.2 and 1.3.2.1 are used to calculate the probability density and cumulative probability distribution of the minimum running depth and the ultimate running depth of the conductor of the target well, as shown in Fig. 1.18.

Select the confidence level $J = J_2 - J_1 = 90\% - 10\% = 80\%$, then in Fig. 1.18b:

(1) The area between points B and C (67.5–87.1 m) is the absolutely safe running depth of the conductor under the condition of an 80% confidence level. The minimum running depth $H_{\min} = 75$ m calculated for the example deepwater well according to the bearing capacity growth coefficient $\alpha_b = 0.1$ belongs to the absolutely safe area and will not cause the subsidence of the subsea wellhead. This is mainly due to the selection of appropriate upper and lower limits of the safety margin when calculating the minimum running depth of the conductor, as shown in Eq. (1.37). However, if the safety margin is too large, the calculated minimum running depth of the conductor will be too large, and there is a risk that the conductor cannot be run in place.

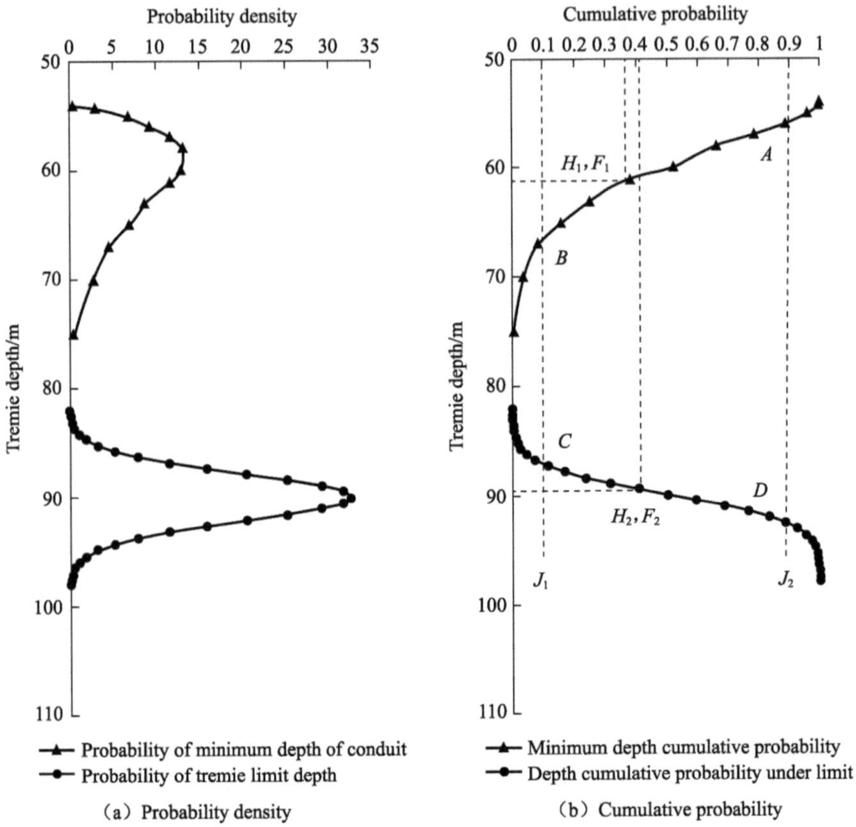

Fig. 1.18 Probability density and cumulative probability distribution of the minimum and ultimate depth of conductor in the target well

(2) If the designed running depth of the conductor is in the area between points A and B (56.3–67.5 m), for example, $H_1 = 61.5$ m, then there is a risk that the subsea wellhead will subside due to the insufficient bearing capacity of the conductor. The specific probability of its occurrence $P_v = 0.36 - 0.1 = 0.26 = 26\%$.
(3) If the designed running depth of the conductor is in the area between points C and D (87.1–88.5 m), for example, $H_2 = 89.5$ m, then there is a risk that the conductor cannot be run in place. The specific probability of its occurrence $P_{\text{fail}} = 0.41 - 0.1 = 0.31 = 31\%$.

In order to improve the longitudinal mechanical stability of the subsea wellhead of the example deepwater well, according to the analysis of the influence of various factors on the characteristic parameters of the longitudinal mechanical stability of the subsea wellhead in Sect. 1.2.2, the following measures can be adopted:

(1) Outer diameter of the conductor: Try to select a conductor with a larger size, which is consistent with improving the lateral mechanical stability of the wellhead.
(2) Recovery time: Appropriately extend the recovery time. Especially when the calculated ultimate running depth of the conductor is small (there is a risk of not being able to be run in place smoothly), the designed running depth of the conductor can be reduced by extending the recovery time.
(3) Bearing capacity growth coefficient: Use the data of the drilled wells in the block to predict the bearing capacity growth coefficient of the target well. The higher the abundance of the block data, the more conducive it is to obtaining an accurate bearing capacity growth coefficient of the target well.

1.5 Summary

(1) The specific types of instability of the mechanical stability of subsea wellheads in deepwater drilling are summarized, namely: wellhead inclination and subsidence. The inclination of the subsea wellhead is mainly due to the deformation of the riser under the actions such as the drift of the drilling vessel and the impact of sea currents, which in turn causes the lateral displacement of the subsea wellhead, and it mostly occurs in the subsequent drilling operations after the cementing of the surface casing; the subsidence of the subsea wellhead is mainly due to the fact that the designed running depth of the conductor is too small, resulting in the bearing capacity of the conductor after it is run in place being insufficient to support the weight of the wellhead device and the subsequent casing, which usually occurs before the cementing of the surface casing.
(2) By analyzing the influence of different factors on the characteristic parameters (lateral displacement and offset angle of the wellhead) of the lateral mechanical stability of the subsea wellhead, it is found that most of the environmental loads and construction parameters have an impact on them to varying degrees. However, the running depth of the conductor (greater than 20 m) has no impact on the lateral mechanical stability of the subsea wellhead, the increment of the underwater unit weight of the soil has basically no impact on the lateral mechanical stability of the subsea wellhead, and the increment of the undrained shear strength of the seabed soil affects the displacement and offset angle of the wellhead within the shallow seabed formation interval of [0, 30]m.
(3) Through sensitivity analysis, according to the relative order of magnitude and size of the sensitive factors $|a_i|$ and the influence baseline b_i, the factors that have a relatively large and sensitive influence on the displacement and offset angle of the subsea wellhead of the example deepwater well are screened out, and the measures to improve the lateral mechanical stability of the subsea wellhead are proposed:

(a) On the basis of ensuring the control of the deformation of the riser and improving its stress state, the top tension of the riser should be reduced as much as possible.
(b) A conductor with a large outer diameter should be selected as much as possible.
(c) The sea current velocity at the water surface is an environmental load and irresistible. The sea current velocity at the water surface should be closely monitored, and other favorable parameters should be adjusted in a timely manner.
(d) The area with a large initial undrained shear strength of the shallow seabed soil should be selected for operation as much as possible.
(e) The drag coefficient is a factor that has a relatively large and sensitive influence on the lateral mechanical stability of the subsea wellhead. This factor is an empirical coefficient, and an appropriate value needs to be selected according to the region and engineering experience.
(f) The average platform drift should be controlled within a small range.
(g) The elastic modulus of the cement sheath is a factor that has a relatively large but not very sensitive influence on the lateral mechanical stability of the subsea wellhead. It is not recommended to use it as the primary adjustment parameter for improving the lateral mechanical stability of the subsea wellhead.
(h) The tidal current velocity at the water surface is a factor that has a relatively large but not very sensitive influence on the lateral mechanical stability of the subsea wellhead. When a_i is positive, it means that the smaller its value is, the more conducive it is to the lateral mechanical stability of the wellhead. However, this factor is an environmental load, and it is necessary to conduct real-time monitoring to adjust other parameters in a timely manner.

(4) Through analysis, the running depth of the conductor $H_{\min}$ when $\Delta Q = \varepsilon$ (ε is the safety margin) is taken as the characteristic parameter of the longitudinal mechanical stability of the subsea wellhead. By analyzing the influence of various factors on the running depth of the conductor and its sensitivity analysis, it is found that the bearing capacity growth coefficient and the recovery time need to be fully considered in the design of the running depth of the conductor. The value of the bearing capacity growth coefficient should be determined as accurately as possible, and the recovery time should be appropriately extended (not less than 1d). At the same time, the outer diameter of the conductor is also a sensitive factor affecting the lateral mechanical stability of the subsea wellhead. Selecting a conductor with a larger outer diameter is conducive to improving the lateral and longitudinal mechanical stability of the subsea wellhead at the same time.
(5) By studying the factors that affect the stability of the subsea wellhead and are uncertain and sensitive, a discrimination method for the instability risk of the subsea wellhead considering uncertain factors is established, and the mechanical

stability of the subsea wellhead of the example deepwater well is discriminated. At the same time, suggestions for improving its stability are given:

(a) The displacement of the subsea wellhead of the example deepwater well basically meets the safety requirements (89%), while there is a relatively high probability (24%) that the offset angle cannot meet the safety requirements. In order to improve its lateral mechanical stability, on the basis of ensuring the control of the deformation of the riser and improving its stress state, the top tension of the riser should be reduced as much as possible; a conductor with a large outer diameter should be selected as much as possible; the area with a large initial undrained shear strength of the shallow seabed soil should be selected for operation as much as possible; during the construction process, the average platform drift should be controlled within a small range, and the drift of the platform needs to be monitored in real time to ensure that the average drift does not exceed $S_0 = 51.86\,\text{m}$; during the construction process, the sea current velocity at the water surface should be closely monitored. Once it is found that $u_w > 1.23\,\text{m/s}$, it is necessary to timely and appropriately reduce the top tension of the riser, and try to control the platform drift as much as possible, etc.

(b) Based on an 80% confidence level, the absolutely safe running depth of the conductor (67.5–87.1 m) of the example deepwater well is determined. In order to improve the longitudinal mechanical stability of the subsea wellhead, a conductor with a larger size should be selected as much as possible, which is consistent with improving the lateral mechanical stability of the wellhead; the recovery time should be appropriately extended. Especially when the calculated ultimate running depth of the conductor is small (there is a risk of not being able to be run in place smoothly), the designed running depth of the conductor can be reduced by extending the recovery time; use the data of the drilled wells in the block to predict the bearing capacity growth coefficient of the target well. The higher the abundance of the block data, the more conducive it is to obtaining an accurate bearing capacity growth coefficient of the target well.

It is worth noting that with the application and development of intelligent algorithms in the petroleum engineering field, the assessment of subsea wellhead instability risks has also evolved towards intelligence, enabling the consideration of more factors to adapt to increasingly complex and harsh environments and working conditions. The fundamental theoretical methods introduced in this chapter can provide effective mechanistic support for data-driven approaches.

Chapter 2
Risk Assessment Method for Well Control During Gas Kick in Deepwater Drilling

Abstract During the drilling process, if a high-pressure gas layer is encountered during drilling, gas kick may occur. In the initial stage of gas kick, it is difficult to accurately detect it through methods or equipment such as pit liquid level detection and flow difference overflow detection. Moreover, as the well depth increases, the difficulty of detecting gas kick in advance increases, and the reaction time left for well control is significantly reduced, making early detection of gas kick a key and difficult point in deep well control. For deepwater drilling, firstly, as the water depth increases, the safety density window narrows, making it more prone to gas kick; Secondly, due to the presence of deep water and high static pressure environments, the overall scale of bubbles reaching the seabed wellhead is still relatively small, resulting in the possibility that the increment of the pit on the sea surface may not reach the warning value when gas enters the riser, making it difficult to detect gas kick in a timely manner through conventional methods. Therefore, the risk of gas kick and well control in deepwater drilling has significantly increased compared to shallow sea drilling. How to timely and accurately detect gas kick and ensure well killing safety has become the top priority of deepwater drilling well control work. On the basis of introducing the calculation model of wellbore annular pressure during gas kick in deepwater drilling, this chapter analyzes the factors that have a significant and uncertain impact on the safety of gas kick and well killing by calculating the risk of well killing under different factor conditions, and establishes a quantitative calculation method for uncertain factors; At the same time, calculate the impact of different factors on gas kick in deepwater drilling, analyze the relationship between the location of gas kick (cross-sectional gas content) and overflow flow (pit increment), explore new ideas for early monitoring of gas kick in deepwater drilling, in order to accurately detect gas kick in advance and quantitatively describe the degree of gas kick.

Y. Xu, *Assessment of Risks Related to Wellbore Safety in Deepwater Drilling*,
https://doi.org/10.1007/978-981-95-1668-1_2

2.1 Calculation Model of Wellbore Annular Pressure Field During Gas Kick in Deepwater Drilling

In recent years, the multiphase flow calculation model in the wellbore annulus has become mature and can basically meet the accuracy requirements of on-site design and construction. However, in deepwater drilling, due to unique environmental factors such as deep water and low temperature, external temperature has a significant impact on the rheological properties, density, and gas–liquid two-phase flow pattern of the drilling fluid in the wellbore annulus. Therefore, it is necessary to improve the calculation accuracy of the Temperature model of seawater section and the wellbore heat transfer model.

2.1.1 *Control Equation*

The temperature model for the deep-water ocean segment is based on a large amount of measured data fitting and has regional characteristics. This book uses the temperature depth model of the southern South China Sea introduced by Yonghai Gao et al. as the external temperature control equation for the seawater section.

For intervals with a water depth greater than 200 m, the following formula is used for calculation:

$$T_{\text{sea}} = a_1 + \frac{a_2}{1 + e^{(h+a_0)/a_3}}, h \geq 200\,\text{m} \tag{2.1}$$

For intervals with a water depth of less than 200 m, different calculation formulas are used according to the seasonal changes in the sea area:

$$\text{Spring: } T_{\text{sea}} = \frac{T_S(200-h) + 13.68h}{200}, 0 \leq h < 200\,\text{m} \tag{2.2}$$

$$\text{Summer: } \begin{cases} T_{\text{sea}} = T_S, 0 \leq h < 20\,\text{m}; \\ T_{\text{sea}} = \frac{T_S(200-h)+13.7(h-20)}{180}, 20 \leq h < 200\,\text{m} \end{cases} \tag{2.3}$$

$$\text{Autumn: } \begin{cases} T_{\text{sea}} = T_S, 0 \leq h < 50\,\text{m}; \\ T_{\text{sea}} = \frac{T_S(200-h)+13.7(h-50)}{150}, 50 \leq h < 200\,\text{m} \end{cases} \tag{2.4}$$

$$\text{Winter: } \begin{cases} T_{\text{sea}} = T_S, 0 \leq h < 100\,\text{m}; \\ T_{\text{sea}} = \frac{T_S(200-h)+13.7(h-100)}{100}, 100 \leq h < 200\,\text{m} \end{cases} \tag{2.5}$$

In the formula, $a_0 = 130.1, a_1 = 2.307, a_2 = 37.1, a_3 = 402.7$. T_{sea} is the temperature of seawater, °C; T_s is the surface temperature of seawater, °C; h is the depth of seawater, m.

The commonly used control equations for the wellbore annular pressure field during gas kick in deepwater drilling are as follows:

(1) Temperature field control equation:

Inside the drill string:

$$\frac{\pi\left[T_a(z,t)-T_p(z,t)\right]}{\frac{1}{h_{pi}d_{pi}}+\frac{1}{h_{po}d_{po}}+\frac{ln\left(\frac{d_{po}}{/d_{pi}}\right)}{2\lambda_{dp}}}-S_pV_p\frac{\partial\left[\rho_p(z,t)c_p(z,t)T_p(z,t)\right]}{\partial z}+Q_{cp}$$
$$=S_p\frac{\partial\left[\rho(z,t)c_p(z,t)T_p(z,t)\right]}{\partial t} \tag{2.6}$$

Formation section annulus:

$$\frac{2\pi\lambda_f d_{co}U_a}{d_{co}U_a f(t)+2\lambda_f}(T_f-T_a)-\frac{\pi\left[T_a(z,t)-T_p(z,t)\right]}{\frac{1}{h_{pi}d_{pi}}+\frac{1}{h_{po}d_{po}}+\frac{\ln\left(d_{po}/d_{pi}\right)}{2\lambda_{dp}}}$$
$$+S_aV_a\frac{\partial[\rho_a(z,t)c_a(z,t)T_a(z,t)]}{\partial z}+Q_{ca}=S_a\frac{\partial[\rho_a(z,t)c_a(z,t)T_a(z,t)]}{\partial t} \tag{2.7}$$

Among them,

$$f(t)=1.1281\sqrt{t_D}\left[1-0.3\sqrt{t_D}\right]10^{-10}\le t_D\le 1.5 \tag{2.8}$$

$$f(t)=(0.4036+0.5\ln t_D)\left(1+\frac{0.6}{t_D}\right)\quad t_D\ge 1.5 \tag{2.9}$$

$$t_D=\frac{4\alpha t}{d_{wb}^2},\ \alpha=\frac{\lambda_f}{\rho_f c_f} \tag{2.10}$$

$$Ua^{-1}=\frac{1}{h_f}+\frac{d_{co}\ln(d_{wb})}{2\lambda_e}\quad \text{(Not considering casing)} \tag{2.11}$$

Seawater section annulus:

$$\frac{\pi[T_s(z,t)-T_a(z,t)]}{\frac{1}{h_{ri}d_{ri}}+\frac{1}{h_{io}d_{io}}+\frac{\ln(d_{ro}/d_{ri})}{2\lambda_r}+\frac{\ln(d_{io}/d_{ii})}{2\lambda_i}}-\frac{\pi\left[T_a(z,t)-T_p(z,t)\right]}{\frac{1}{h_{pi}d_{pi}}+\frac{1}{h_{po}d_{po}}+\frac{\ln\left(d_{po}/d_{pi}\right)}{2\lambda_{dp}}}+$$
$$S_aV_a\frac{\partial[\rho_a(z,t)c_a(z,t)T_a(z,t)]}{\partial z}+Q_{ca}=S_a\frac{\partial[\rho_a(z,t)c_a(z,t)T_a(z,t)]}{\partial t} \tag{2.12}$$

Boundary conditions:
Entrance: $T_p(z=0,t)=\text{constant}$;
Bottom of the well: $T_p(z=H,t)=T_a(z=H,t)$;
Stratum: $T_f(z,r\to\infty,t)=T_f(z,r\to\infty,t=0)$.

(2) Control equation for pressure field of gas–liquid two-phase flow:

Gas phase continuity equation:

$$\frac{\partial\left(S_a\rho_g E_g\right)}{\partial t}+\frac{\partial\left(S_a\rho_g E_g V_g\right)}{\partial z}=Q_g \tag{2.13}$$

Liquid phase continuity equation;

$$\frac{\partial\left[S_a\rho_l\left(1-E_g\right)\right]}{\partial t}+\frac{\partial\left[S_a\rho_l\left(1-E_g\right)V_l\right]}{\partial z}=Q_l \tag{2.14}$$

Gas liquid two-phase mixing equation:

$$\frac{\partial\left[S_a\rho_g E_g V_g+S_a\rho_l\left(1-E_g\right)V_l\right]}{\partial t}+\frac{\partial\left[S_a\rho_g E_g V_g^2+S_a\rho_l\left(1-E_g\right)V_l^2\right]}{\partial z}+ \\ S_a\frac{\partial P}{\partial z}+S_a\left(\frac{\partial P}{\partial z}\right)_{\text{fric}}+S_a\left[\rho_g E_g+\rho_l\left(1-E_g\right)\right]g=0 \tag{2.15}$$

Empirical equation for gas rise rate:

$$V_g=C_0\left[V_g E_g+V_l E_l\right]+v_{rg} \tag{2.16}$$

Gas phase state equation:

$$\rho_g=3486.6\frac{P\rho_{gs}}{Z(T+273)} \tag{2.17}$$

Boundary conditions:

$$\begin{cases} Q_g(t,0)=Q_g \\ Q_l(t,0)=Q_l \\ P(0,N)=\text{pressure atmosphere} \end{cases} \tag{2.18}$$

Among the above types, S_a is the cross-sectional area of the annulus, m^2; The volume fractions of gas and liquid phases, dimensionless; Gas phase density, gas phase relative density, and drilling fluid density are respectively E_g, E_l; The velocities of the gas phase and liquid phase are respectively ρ_g, ρ_{gs}, ρ_l, kg/m^3; The velocities of the gas and liquid phases are respectively v_g, v_l. The mass of gas produced by a gas layer per unit time and thickness are Q_g, $kg/(s\cdot m)$; v_{lo} is the fluid velocity at the wellhead; Q_{lo} is the pump flow rate, m^3/s; For frictional pressure drop along the way is $(\frac{\partial P}{\partial z})_{\text{fric}}$, Pa/m; $\frac{\partial P}{\partial z}$ is the pressure drop of the liquid column, Pa/m; g is the gravitational acceleration, m^2/s; C_0 is the gas phase distribution coefficient, dimensionless; v_{rg} is the gas slip velocity, m/s; Z is the gas compression factor, dimensionless; T Is the temperature, °C; P is pressure, Pa; $P(t, \mathrm{N})$ is the wellhead casing pressure, MPa.

2.1.2 Discretization Processing

The above control equation system is relatively complex and difficult to calculate through analytical algorithms. Numerical methods can be used to discretize it, resulting in the following numerical equation system:

(1) Discretization equation system for temperature field:

Inside the drill string:

$$\begin{aligned}&\frac{\pi\left[T_{\mathrm{a}}(i,k)-T_{\mathrm{p}}(i,k)\right]}{\frac{1}{h_{\mathrm{pi}}(i,k)d_{\mathrm{pi}}}+\frac{1}{h_{\mathrm{po}}(i,k)d_{\mathrm{po}}}+\frac{\ln\left(d_{\mathrm{po}}\right)}{2\lambda_{dp}}}-S_{\mathrm{p}}V_{\mathrm{p}}\rho_{\mathrm{p}}(i,k)c_{\mathrm{p}}(i,k)\frac{T_{\mathrm{p}}(i,k)-T_{\mathrm{p}}(i,k-1)}{\Delta z}\\&+Q_{\mathrm{cp}}(i,k)=S_{\mathrm{p}}\rho_{\mathrm{p}}(i,k)c_{\mathrm{p}}(i,k)\frac{T_{\mathrm{p}}(i,k)-T_{\mathrm{p}}(i-1,k)}{\Delta t}\end{aligned} \tag{2.19}$$

Formation section annulus:

$$\begin{aligned}&\frac{2\pi\lambda_{\mathrm{f}}d_{\mathrm{co}}U_{\mathrm{a}}}{d_{\mathrm{co}}U_{\mathrm{a}}f(t)+2\lambda_{\mathrm{f}}}\left[T_{\mathrm{f}}(i,k)-T_{\mathrm{a}}(i,k)\right]-\frac{\pi\left[\mathrm{T_a}(i,k)-\mathrm{T_p}(i,k)\right]}{\frac{1}{\mathrm{h_{pi}}(i,k)\mathrm{d_{pi}}}+\frac{1}{h_{\mathrm{po}}(i,k)d_{\mathrm{po}}}+\frac{\ln\left(d_{\mathrm{po}}/d_{\mathrm{pi}}\right)}{2\lambda_{\mathrm{dp}}}}\\&+S_{\mathrm{a}}V_{\mathrm{a}}\rho_{\mathrm{a}}(i,k)c_{\mathrm{a}}(i,k)\frac{T_{\mathrm{a}}(i,k+1)-T_{\mathrm{a}}(i,k)}{\Delta z}+Q_{\mathrm{ca}}(i,k)\\&=S_{\mathrm{a}}\rho_{\mathrm{a}}(i,k)c_{\mathrm{a}}(i,k)\frac{T_{\mathrm{a}}(i,k)-T_{\mathrm{a}}(i-1,k)}{\Delta t}\end{aligned} \tag{2.20}$$

Seawater section annulus:

$$\begin{aligned}&\frac{\pi\left[T_{\mathrm{s}}(i,k)-T_{\mathrm{a}}(i,k)\right]}{\frac{1}{h_{\mathrm{ri}}(i,k)d_{\mathrm{ri}}}+\frac{1}{h_{\mathrm{ro}}(i,k)d_{\mathrm{ro}}}+\frac{\ln(d_{\mathrm{ro}}/d_{\mathrm{ri}})}{2\lambda_{\mathrm{r}}}+\frac{\ln(d_{\mathrm{io}}/d_{\mathrm{ii}})}{2\lambda_{\mathrm{i}}}}-\frac{\pi\left[T_{\mathrm{a}}(i,k)-T_{\mathrm{p}}(i,k)\right]}{\frac{1}{\mathrm{h_{pi}}(i,k)\mathrm{d_{pi}}}+\frac{1}{h_{po}(\mathrm{i,k})d_{po}}+\frac{\ln\left(d_{\mathrm{po}}/d_{\mathrm{pi}}\right)}{2\lambda_{\mathrm{dp}}}}\\&+S_{\mathrm{a}}V_{\mathrm{a}}\rho_{\mathrm{a}}(i,k)c_{\mathrm{a}}(i,k)\frac{T_{\mathrm{a}}(i,k+1)-T_{\mathrm{a}}(i,k)}{\Delta z}\\&+Q_{\mathrm{ca}}(i,k)=S_{\mathrm{a}}\rho_{\mathrm{a}}(i,k)c_{\mathrm{a}}(i,k)\frac{T_{\mathrm{a}}(i,k)-T_{\mathrm{a}}(i-1,k)}{\Delta t}\end{aligned} \tag{2.21}$$

(2) Differential equation system for pressure field of gas–liquid two-phase flow:

Gas phase continuity equation:

$$(S_{\mathrm{a}}V_{\mathrm{g}}E_{\mathrm{g}}\rho_{\mathrm{g}})_{j+1}^{n+1}-(S_{\mathrm{a}}V_{\mathrm{g}}E_{\mathrm{g}}\rho_{\mathrm{g}})_{j}^{n+1}$$

$$= \frac{\Delta z}{2\Delta t}\left[\left(S_a E_g \rho_g\right)_j^n + \left(S_a E_g \rho_g\right)_{j+1}^n - \left(S_a E_g \rho_g\right)_j^{n+1} + \left(S_a E_g \rho_g\right)_{j+1}^{n+1}\right] \tag{2.22}$$

Liquid phase continuity equation:

$$\begin{aligned}
&(S_a V_l E_l \rho_l)_{j+1}^{n+1} - (S_a V_l E_l \rho_l)_j^{n+1} \\
&= \frac{\Delta \tau}{2\Delta t}\left[(S_a E_l \rho_l)_j^n + (S_a E_l \rho_l)_{j+1}^n - (S_a E_l \rho_l)_j^{n+1} + (S_a E_l \rho_l)_{j+1}^{n+1}\right]
\end{aligned} \tag{2.23}$$

Gas liquid two-phase mixing equation:

$$\begin{aligned}
P_{j+1}^{n+1} - P_j^{n+1} = {} & \frac{\Delta z}{2\Delta t}\Big[\left(\rho_g E_g V_g + \rho_l E_l V_l\right)_j^n + \left(\rho_g E_g V_g + \rho_l E_l V_l\right)_{j+1}^n \\
& -\left(\rho_g E_g V_g + \rho_l E_l V_l\right)_j^{n+1} - \left(\rho_g E_g V_g + \rho_l E_l V_l\right)_{j+1}^{n+1}\Big] \\
& + \left[\left(\rho_g E_g V_g^2 + \rho_l E_l V_l^2\right)_j^{n+1} - \left(\rho_g E_g V_g^2 + \rho_l E_l V_l^2\right)_{j+1}^{n+1}\right] \\
& - \frac{\Delta z}{2}\left\{\left[\left(\rho_g E_g g + \rho_l E_l g\right]_j^{n+1} + \left[\rho_g E_g g + \rho_l E_l g\right]_{j+1}^{n+1}\right\} \\
& - \frac{\Delta z}{2}\left\{\left[\left(\frac{\partial P}{\partial z}\right)_{\text{fric}}\right]_j^{n+1} + \left[\left(\frac{\partial P}{\partial z}\right)_{\text{fric}}\right]_{j+1}^{n+1}\right\}
\end{aligned} \tag{2.24}$$

Thus, numerical solutions can be obtained for the control equations of temperature and pressure fields based on their respective boundary conditions.

2.2 Analysis of the Influence of Different Factors on the Gas Kick and Overflow Law in Deepwater Drilling

The environmental factors that affect the gas kick process in deep water drilling mainly include water depth, depth below the mud line, formation pore pressure, and formation permeability. Taking a deep water well in the South China Sea as an example, using the calculation model of the pressure field in the wellbore annulus during gas kick in deepwater drilling in 2.1, the relationship between gas arrival position and overflow rate is calculated. At the same time, the distribution of gas content in the cross-section of the wellbore annulus when gas reaches the seabed wellhead is calculated, as shown in Figs. 2.1 and 2.2. The specific parameters of this deep water well are as follows:

The water depth is 2250 m, the well depth is 3803 m, the cemented section is 1200 m, the open-hole section is 353 m, the seawater surface temperature is 15 °C, the geothermal gradient is $P_{ci}(i = \text{casing level})$, the outer diameter of the riser is 508 mm, the circulating flowrate is 46 L/s, the drilling fluid density is 1.21 g/cm^3, the drilling fluid thermal conductivity coefficient is 1.72 W/(m ·° C), the viscosity coefficient is

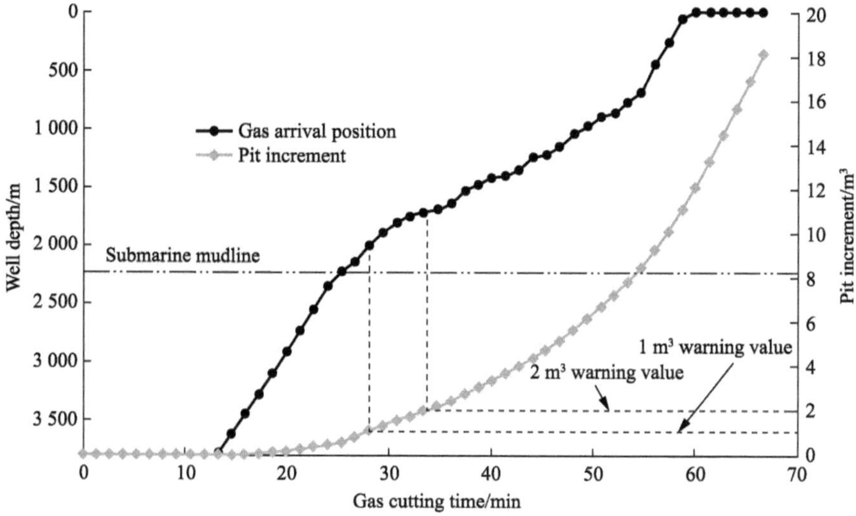

Fig. 2.1 The gas position in wellbore and increment of mud pool after gas cut

Fig. 2.2 The gas void fraction distribution in wellbore when the gas cut reaches subsea wellhead

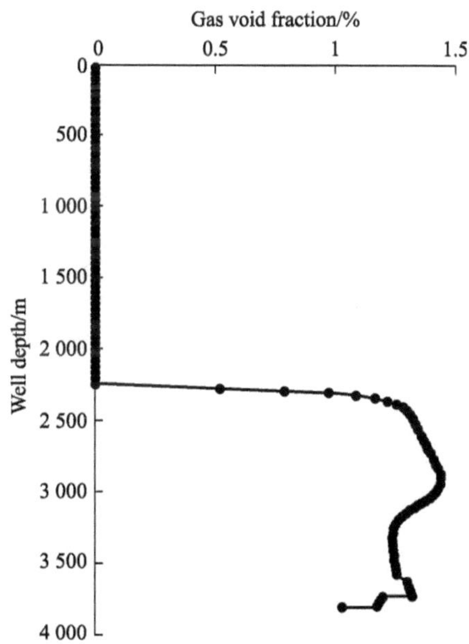

0.25, the fluidity index is 0.38, the formation porosity is 30%, the permeability is 10 md, the formation thermal conductivity coefficient is 2.21 W/(m ·° C), and the original formation pressure is 51.2 MPa.

From Fig. 2.1, it can be seen that after the gas kick, the increment of the mud pit near the seabed wellhead has not yet reached the warning value of 1 m^3 when the gas reaches it. However, when the increment of the mud pit reaches 1 m^3, the gas has been transported upward in the riser for nearly 3 min, rising by 234 m; When the mud pit increment reached 2 m^3, the gas had been migrating upwards in the riser for nearly 9 min, rising by 558 m. It can be seen that following the warning value of 1 m^3 for well control operations will increase the risk of well control. At the same time, as shown in Fig. 3.2, when the gas reaches the seabed wellhead, the cross-sectional gas content of the early gas is relatively small, indicating that there is less gas entering the wellbore in the early stage. If gas kick can be detected at this time and well control operations can be carried out, the risk of well control can be reduced. Therefore, calculating the overflow rate of gas reaching the seabed wellhead under different conditions and the gas arrival positions at overflow rates of 1 m^3 and 2 m^3 can help analyze the development law of gas kick under different conditions and well control risks.

(1) The impact of different water depths

Due to changes in water depth affecting the magnitude of bottom hole pressure and formation pressure, in order to ensure comparability of the calculation results, the bottom hole pressure difference under different water depth conditions is kept constant (1 MPa), that is, the pore pressure of the formation is adjusted accordingly based on changes in water depth. Calculate the overflow rate of gas reaching the seabed wellhead after gas kick at depths of 800 m, 1525 m, 2250 m, and 2975 m, as well as the gas arrival positions at overflow rates of 1 m^3 and 2 m^3 as shown in Figs. 2.3 and 2.4.

From Fig. 2.3, it can be seen that the overflow rate of gas reaching the seabed wellhead decreases with the increase of water depth, but the change amplitude is small. Moreover, the overflow rate under the calculated water depth (800–2975 m) is all less than 1 m^3. This is because in deep water and high static pressure environments, the overall scale of gas kick before reaching the riser is small, which is not easy

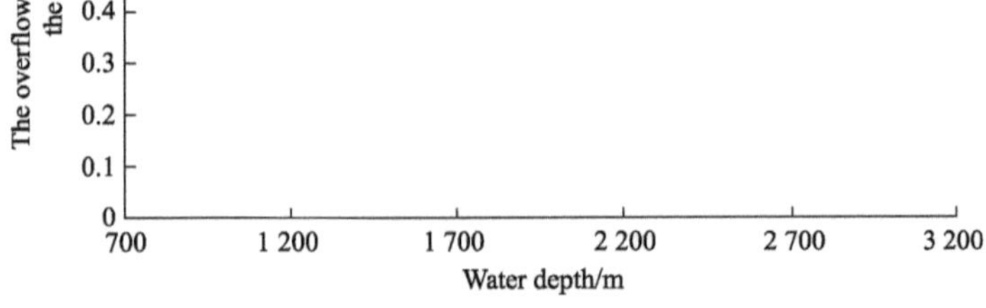

Fig. 2.3 The overflow value when gas cut reaches subsea wellhead under different water depths

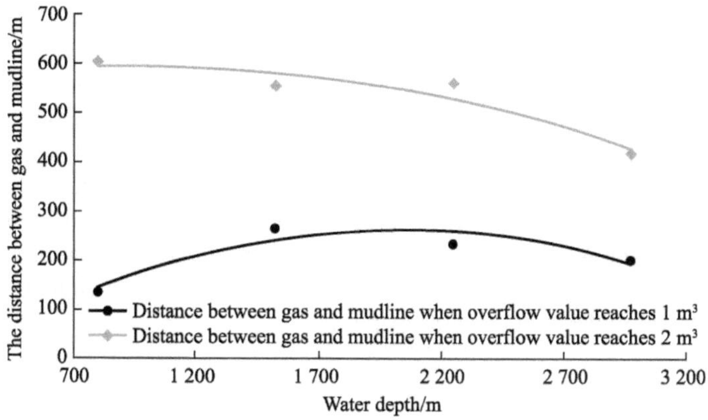

Fig. 2.4 The distance of gas position from mud line when overflow value reaches 1 m^3 or 2 m^3 under different water depths

to cause an increase in pit increment; The deeper the water depth, the greater the compression effect of high static pressure on gas in deep water. However, under deep water conditions, the overall scale of gas has been sufficiently compressed, so the overflow rate of gas when it reaches the seabed wellhead decreases with increasing water depth, but the change amplitude is relatively small.

From Fig. 2.4, it can be seen that the distance between the gas and the seabed mudline (positive above the mudline and negative below) at an overflow rate of 1 m^3 increases first and then decreases with the increase of water depth. However, the distance between the gas and the seabed mudline at an overflow rate of 2 m^3 decreases with the increase of water depth, and the trend of change is not consistent. This is because the deeper the water depth, the greater the pressure effect of high static pressure on the gas in deep water, After the gas kick enters the riser, the flow pattern is more prone to change due to the change in the horizontal size of the wellbore annulus (from bubbly flow to slug flow, etc.), resulting in a decrease in the upward velocity of the gas kick in the riser.

It can be seen that deep water drilling with a larger depth will result in a mud pit increment of less than 1 m^3 when gas reaches the seabed wellhead, and through the mud pit increment method, it is found that a large amount of gas kick gas has accumulated in the riser during gas kick, leading to an increased risk of well control.

(2) The influence of depths below different mud lines

Due to changes in the depth below the mud line (well depth), it can also affect the magnitude of bottom hole pressure and formation pressure. In order to ensure comparability of the calculation results, the bottom hole pressure difference under different depths below the mud line is kept constant (1 MPa), that is, the pore pressure of the formation is adjusted accordingly based on changes in well depth. Calculate the overflow rate of gas reaching the seabed wellhead after gas kick at depths of

900 m, 1553 m, 2100 m, 2700 m, and 3300 m below the mud line, as well as the gas arrival positions at overflow rates of 1 m^3 and 2 m^3, as shown in Figs. 2.5 and 2.6.

From Fig. 2.5, it can be seen that the overflow rate of gas reaching the seabed wellhead increases with the depth below the mudline. When the depth below the mudline is less than 2800 m, the overflow rate is less than 1 m^3. This is because, when the water depth is constant, the greater the depth below the mud line, the longer the migration path of gas kick gas below the mud line, and the more gas accumulates in the wellbore. Although there is a compression effect of deep water high static pressure environment, the accumulation of small amounts of gas will gradually increase the increment of the pit.

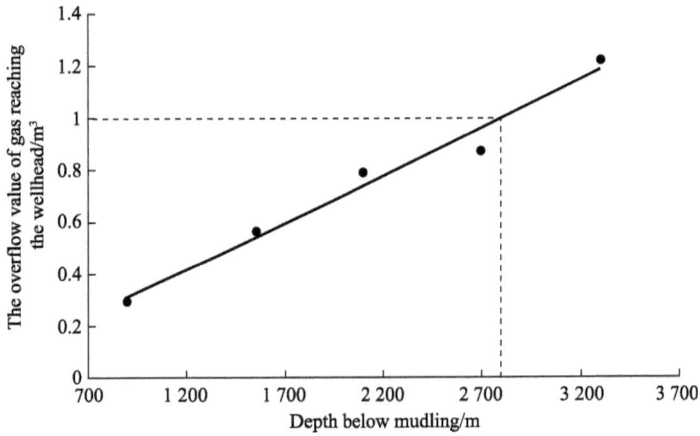

Fig. 2.5 The overflow value when gas cut reaches subsea wellhead under different well depths

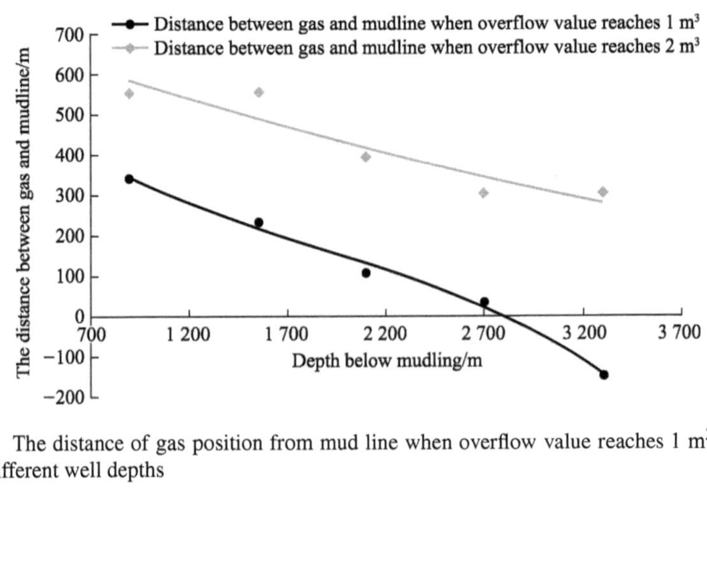

Fig. 2.6 The distance of gas position from mud line when overflow value reaches 1 m^3 or 2 m^3 under different well depths

From Fig. 2.6, it can be seen that the distance between the gas at overflow rates of 1 m^3 and 2 m^3 and the seabed mudline (positive above the mudline and negative below) decreases with increasing well depth. When the depth below the mudline is greater than 2800 m, the gas at overflow rate of 1 m^3 has not yet reached the seabed wellhead, but the gas at overflow rate of 2 m^3 has already migrated hundreds of meters upwards in the riser.

From this, it can be seen that when the water depth is 2250 m, for deep-water drilling with a depth below the mud line of less than about 2800 m, the smaller the depth below the mud line, the more unfavorable the existing mud pool liquid level monitoring method is for monitoring gas kick in deep-water drilling. Shallow gas is generally buried within a range of 1500 m below the mud line, so effective prediction and control of gas kick risk caused by shallow gas and other factors are necessary in deepwater drilling.

(3) The influence of different bottomhole pressure differences

Calculate the overflow rate of gas reaching the seabed wellhead after gas kick under the conditions of bottom hole pressure difference of 0.5 MPa, 1 MPa, 1.5 MPa, 2 MPa, and 4.5 MPa, as well as the gas arrival position at overflow rates of 1 m^3 and 2 m^3, as shown in Figs. 2.7 and 2.8.

As shown in Fig. 2.7, the overflow rate of gas reaching the seabed wellhead increases with the increase of bottomhole pressure difference. When the bottomhole pressure difference is less than 4.3 MPa, the overflow rate is less than 1 m^3. This is because the greater the pressure difference at the bottom of the well, the faster the formation gas enters the wellbore annulus, resulting in more gas invading the wellbore when the gas reaches the seabed wellhead, which is more likely to cause an increase in the overflow flow rate of the mud pool.

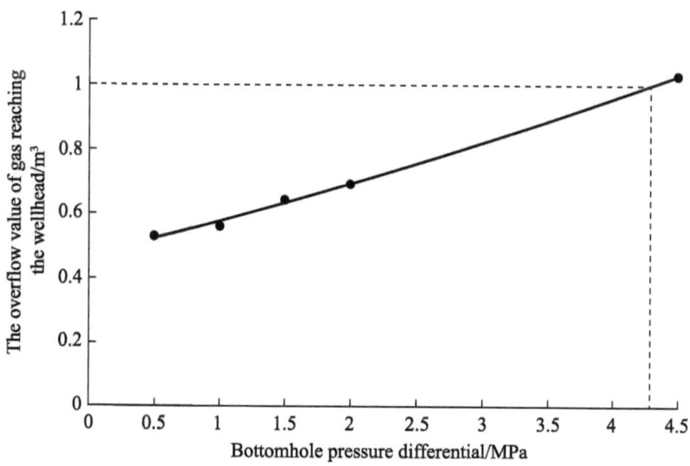

Fig. 2.7 The overflow value when gas cut reaches subsea wellhead under different bottom hole pressure differential

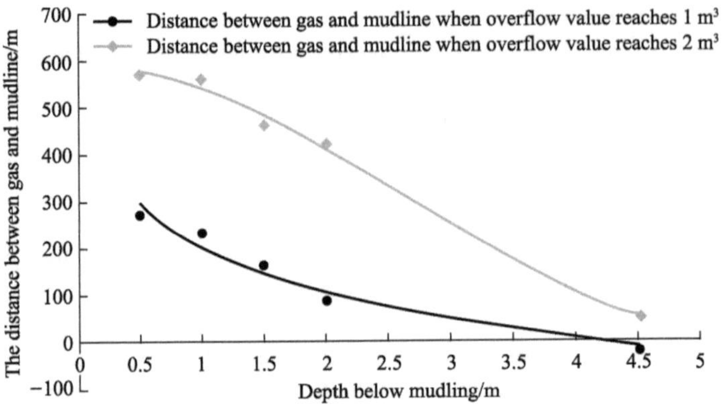

Fig. 2.8 The distance of gas position from mud line when overflow value reaches 1 m^3 or 2 m^3 under different bottom hole pressure differential

As shown in Fig. 2.8, the distance between the gas and the seabed mud line (positive above the mud line and negative below) decreases with the increase of the bottomhole pressure difference at overflow rates of 1 m^3 and 2 m^3. This is because the smaller the pressure difference at the bottom of the well, the slower the formation gas enters the wellbore annulus, resulting in a longer time for the accumulated gas kick volume in the wellbore annulus to reach 1 m^3 and 2 m^3, that is, the longer the gas kick gas migrates in the riser.

It can be seen that for deep water drilling, the magnitude of the bottom hole pressure difference directly affects the overflow rate of gas when it reaches the seabed wellhead and the distance of gas migration in the riser when the mud pool increment reaches 1 m^3 and 2 m^3. Therefore, accurate prediction of formation pressure is crucial for controlling gas kick in deep water drilling.

(4) The influence of different formation permeability

The upper part of the reservoir in deep water drilling wells is mainly composed of shale, and the permeability of high-pressure gas reservoirs is generally lower than 50 md. Therefore, the overflow rate of gas reaching the seabed wellhead after gas kick under the conditions of permeability of 1 md, 5 md, 10 md, 20 md, and 50 md, as well as the gas arrival position at overflow rates of 1 m^3 and 2 m^3, are calculated separately, as shown in Figs. 2.9 and 2.10.

From Fig. 2.9, it can be seen that the overflow rate of gas reaching the seabed wellhead increases with the increase of layer permeability. When the permeability is less than 44 md, the overflow rate remains below 1 m^3. This is because the lower the permeability of the formation, the slower the rate of formation gas entering the wellbore, and the less gas accumulates in the wellbore at the same time, resulting in a smaller overflow rate when the gas reaches the seabed wellhead.

From Fig. 2.10, it can be seen that the distance between the gas and the seabed mud line (positive above the mud line and negative below) increases with the decrease of reservoir permeability when the overflow flow rates are 1 m^3 and 2 m^3. This is

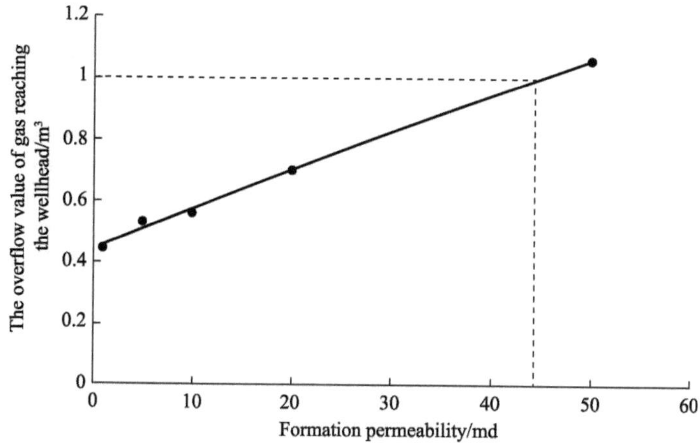

Fig. 2.9 The overflow value when gas cut reaches subsea wellhead under different formation permeability

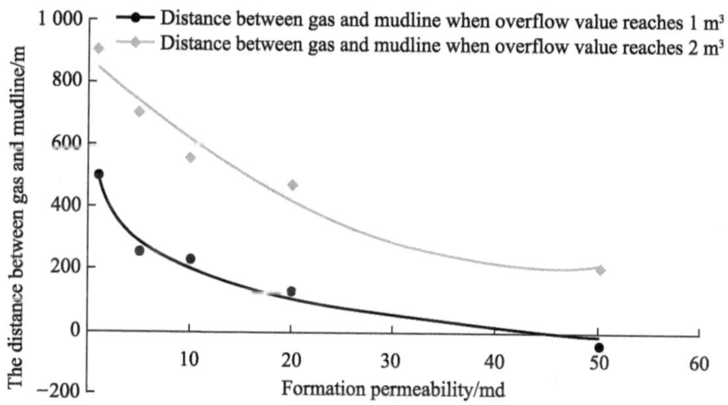

Fig. 2.10 The distance of gas position from mud line when overflow value reaches 1 m^3 or 2 m^3 under different formation permeability

because the lower the formation permeability, the slower the rate of formation gas kick into the wellbore, and the longer it takes for the gas volume in the wellbore to accumulate to 1 m^3 and 2 m^3, resulting in an increase in gas migration time in the riser.

It can be seen that the permeability of the formation has a significant impact on gas kick in deep water drilling. The smaller the permeability of the formation, the more unfavorable it is for the mud pool liquid level monitoring method to monitor gas kick in deep water drilling.

In summary, by analyzing the overflow rate of gas reaching the seabed wellhead after gas kick under different water depths, depths below the mud line, bottom hole pressure difference, and formation permeability conditions, as well as the gas arrival positions at overflow rates of 1 m^3 and 2 m^3, the following conclusions can be drawn:

(1) For deep water drilling (with a depth greater than 800 m), the high static pressure environment in deep water can result in a small increase in mud pool increment when gas reaches the seabed wellhead, and in most cases, it cannot reach the 1 m^3 warning value specified in the offshore drilling manual;
(2) The smaller the depth below the mud line, the more unfavorable the existing mud pool liquid level monitoring method is for monitoring gas kick in deepwater drilling. If shallow gas reservoirs (buried within 1500 m below the mud line) are encountered during deepwater drilling, it is difficult to detect gas kick in a timely manner using the existing mud pool liquid level monitoring method;
(3) For deepwater drilling, the magnitude of the bottomhole pressure difference directly affects the overflow rate of gas when it reaches the seabed wellhead and the distance of gas migration in the riser when the mud pool increment reaches 1 m^3 and 2 m^3. Therefore, accurate prediction of formation pressure is crucial for controlling gas kick in deepwater drilling;
(4) The permeability of the formation has a significant impact on gas kick in deepwater drilling. The lower the permeability of the formation, the less conducive it is to the monitoring of gas kick in deepwater drilling using the mud pool liquid level monitoring method.

2.3 Analysis of Risk Factors and Influencing Laws of Gas Kick and Well Killing

The previous analysis analyzed the relationship between the gas arrival location and overflow rate after gas kick in deepwater drilling, providing a comprehensive understanding of the gas migration and other laws after gas kick. The safety of well killing after gas kick is of utmost importance in well control work. Therefore, it is necessary to analyze the risk factors and influencing laws of gas kick and well killing safety in deepwater drilling.

The main factors that lead to well killing risks in deepwater drilling include errors in overflow monitoring (instrument or human monitoring errors), prediction errors in formation pressure, insufficient pressure bearing capacity of well control devices and wellbore structures, and unreasonable well killing construction parameters. Specifically, it mainly includes: overflow rate, formation pressure, maximum allowable shut in casing pressure, kill fluid density and displacement, etc. Among them, the overflow rate refers to the increase in mud pool volume during well killing measures after gas kick, which is generally monitored by the mud pool liquid level method. According to the provisions of the Marine Drilling Manual, a fixed value of 1 m^3 is taken as an empirical warning value; Formation pressure is predicted by neighboring well logging data or regional seismic data, and its accuracy depends on the accuracy of the prediction method; The maximum allowable shut in casing pressure is calculated based on the pressure bearing capacity of the well control equipment and wellbore structure, and can be accurately determined through calculation; The density and displacement of well killing fluid belong to the parameters of well killing construction, and their values need to be calculated based on drilling conditions and well killing requirements.

Therefore, first introduce the method for determining the maximum allowable shut in casing pressure and kill fluid density.

(1) Upper and lower limits of well killing fluid density

The density of well killing fluid is a key parameter for the success of well killing. If the density is too small, it can lead to well killing failure; If the density is too high, it can lead to underground accidents such as pressure leakage in the formation. Therefore, there are upper and lower limits to the design value of kill fluid density. Through research, the calculation formulas for the upper and lower limits of kill fluid density are as follows:

$$\text{lower limit: } \rho_{\text{k1}} = \rho_{\text{d}} + \frac{P_{\text{sp}}}{0.00981H} + S_{\text{f}} \tag{2.25}$$

$$\text{upper limit: } \rho_{\text{k2}} = \rho_{\text{f}} - S_{\text{b}} - S_{\text{g}} - S_{\text{f}} \tag{2.26}$$

where, ρ_{d} is the density of drilling fluid, g/cm^3; P_{sp} is the vertical pressure for well closure, MPa; S_{h}, S_{b}, S_{g} and are respectively the formation fracture pressure coefficient, suction pressure coefficient, and excitation pressure coefficient, g/cm^3.

The shut in riser pressure in Eq. (2.25) represents the difference between formation pressure and liquid column pressure in the drill string, and can be accurately determined during well killing according to the methods specified in the Marine Drilling Manual. However, before drilling, it is difficult to determine the shut-in pressure when the bottom hole pressure and formation pressure are perfectly balanced due to the lack of knowledge about the changes in shut-in casing pressure and vertical pressure. A preliminary estimate can be made using the following equation.

$$P_{\text{sp}} = 0.00981H \cdot (\rho_{\text{dPrediction Error}} - \Delta\rho - S_{\text{b}}) \tag{2.27}$$

where, $\rho_{\text{dPrediction Error}}$ is the prediction error of formation pressure, g/cm^3; $\Delta\rho$ is additional drilling fluid density, g/cm^3.

(2) Maximum allowable shut in casing pressure

The maximum allowable shut-in casing pressure curve is the maximum shut-in casing pressure allowed under different drilling fluid densities during shut-in, reflecting the limit of wellbore pressure bearing capacity when using drilling hydraulic wells with different densities after overflow occurs. According to the regulations of the offshore drilling manual, the maximum allowable shut-in casing pressure shall not exceed the rated working pressure of the wellhead device the minimum value of 80% of the internal pressure resistance of the casing and the allowable shut in casing pressure for weak formation fracturing pressure. Therefore, the maximum allowable shut in casing pressure curve can be calculated according to the following steps:

(1) Determine the rated working pressure P_{wh} of the wellhead device selected based on the drilling design;
(2) Determine the internal pressure resistance strength $P_{ci}(i = \text{casing level})$ for different openings based on the type of casing used in the cemented section. If there are two or more layers of casing in the wellbore annulus (the tail pipe is not connected back to the wellhead), select the minimum value of the internal pressure resistance strength of each layer of casing;
(3) Based on the wellbore structure and formation fracture pressure profile, determine the minimum value P_{fmin} of formation fracture pressure in the open hole section and its depth H;
(4) Based on the density of drilling fluid reserved for different well killing operations in drilling design, determine the density of different well killing fluids using formula (3.27) the maximum allowable shut in casing pressure ρ_j under the condition of P_j.

$$P_j = min\{P_{\text{wh}}, min(0.8 \cdot P_{\text{ci}}), (P_{\text{fmin}} - \rho_j) \cdot 0.00981 \cdot H\} \tag{2.28}$$

(5) draw $\rho_j - P_j$ curve is used to obtain the maximum allowable shut in casing pressure curve.

From this, the impact of various factors on the risk of gas kick and well killing can be calculated and analyzed. Taking a deep water well in the South China Sea as an example, the well killing risk under different overflow rates, formation pressures, and well killing displacement conditions can be calculated. The main parameters of the well are as follows Table 2.1.

Firstly, according to Eqs. (2.25) and (2.26), the safe kill fluid density range is calculated to be 1.26–1.37 g/cm^3. It is determined that the designed kill fluid density of 1.29 g/cm^3 meets the safety requirements. Then, based on Eq. (2.27), the maximum allowable shut in casing pressure under the designed kill fluid density condition is calculated to be 9.3 MPa. The impact of overflow rate, formation pressure, and well killing displacement on well killing risk is shown in Figs. 2.11, 2.12, and 2.13.

Table 2.1 The main parameters of deepwater well

Parameter	Price	Parameter	Price
Well depth/m	3630	Water depth/m	1298
Inner diameter of riser/mm	482.6	Outer diameter of riser/mm	533.4
Cycle displacement/(L/s)	27.1	Speed/rpm	80
Initial viscosity/Pa.s	55	Formation Pressure/MPa	40.9
Drilling fluid density/(g/cm^3)	1.14	Temperature gradient/(°C/m)	0.0463
Design well killing fluid density/(g/cm^3)	1.29	Rated pressure of blowout preventer/MPa	103

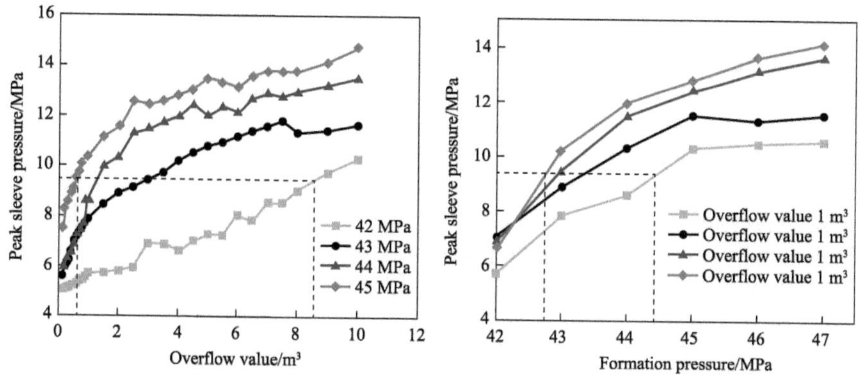

Fig. 2.11 The killing risk with different overflow and formation pressure

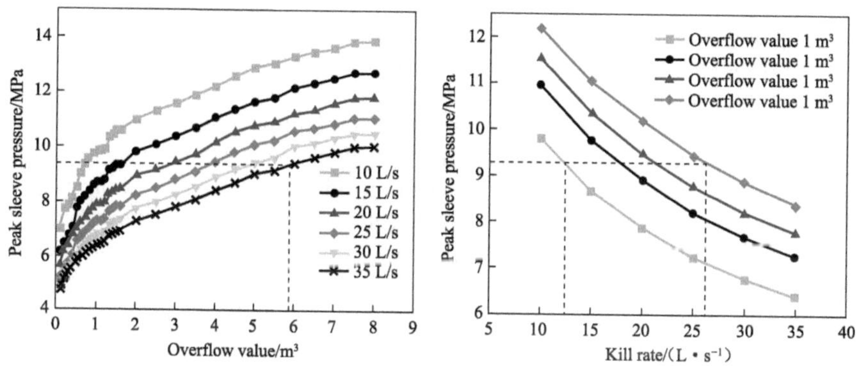

Fig. 2.12 The killing risk with different overflow and kill rate

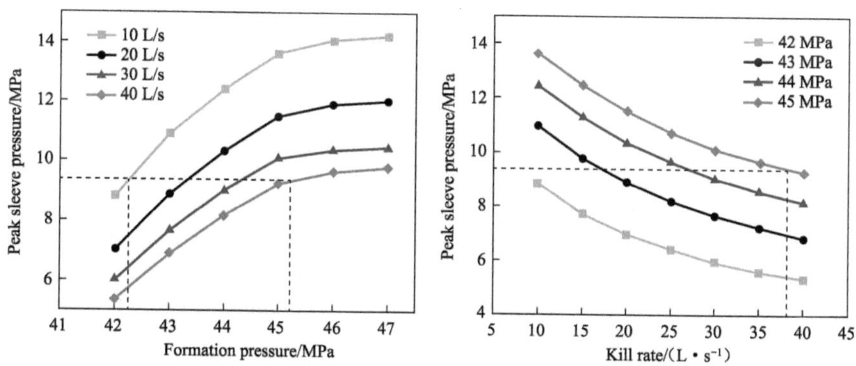

Fig. 2.13 The killing risk with different formation pressure and kill rate

From the above figures, it can be seen that the peak casing pressure after well killing varies greatly depending on the overflow rate, formation pressure, and well killing displacement. The red dashed line in the figure represents the maximum allowable shut in casing pressure. If the peak casing pressure exceeds this value, it is considered to be at risk of damaging the wellhead, casing, or leaking the formation. It can be seen that these three factors have a significant impact on the risk of gas kick and well pressure. At the same time, when the overflow rate is 1 m^3 for well killing, the maximum allowable formation pressure is about 44.5 MPa, and the minimum allowable well killing displacement is 12.5 L/s. Therefore, in order to ensure well killing safety, the well killing displacement should be greater than 12.5 L/s; If the prediction error of formation pressure is large, resulting in an actual formation pressure greater than 44.5 MPa, using the existing 1 m^3 overflow warning value for well killing will inevitably cause the peak pressure of the kill casing to be greater than the maximum allowable shut in casing pressure, thereby increasing the risk of well killing failure.

From the above analysis, it can be concluded that the well killing displacement is a construction parameter, and a reasonable range can be selected for precise control based on the simulation results of gas kick and well killing; The 1 m^3 overflow rate is an empirical warning value, which is an uncertain parameter. For formations with high formation pressure, it may not meet the requirements of well killing safety. Therefore, it is necessary to seek an accurate calculation method for this value; The prediction of formation pressure has errors, and its value has a significant impact on the peak casing pressure (well killing risk). Therefore, when calculating the overflow warning value, the influence of formation pressure prediction errors should be considered.

2.4 Design Method for Overflow Warning Values Applicable to Gas Kick Control in Deepwater Drilling

At present, domestic and foreign scholars have conducted extensive research on the early monitoring of overflow in deepwater drilling, and have achieved rich results. The development of new overflow monitoring equipment and sensors has improved the accuracy of deepwater drilling overflow monitoring, providing a solid foundation for early monitoring of overflow. However, many current studies have focused on the development of overflow monitoring methods or equipment. There is relatively little research on the mud pit incremental method, which is simple to operate, convenient and practical, and widely used in deepwater drilling overflow monitoring. It is generally believed that this method has low accuracy and precision, and is no longer suitable for overflow monitoring in deepwater drilling.

At present, the warning value of the mud pit increment method used in deepwater drilling is still an empirical fixed value, which is an important factor leading to the low accuracy and timeliness of this method. If the overflow warning value applicable to the well or block can be designed based on on-site actual calculation, it can timely and effectively warn of high-risk overflow or gas kick, which will undoubtedly greatly improve the efficiency of deepwater drilling overflow monitoring and reduce the cost of overflow monitoring.

2.4.1 Limitation Analysis of Existing Mud Pit Increment Method in Deepwater Drilling

According to the analysis of the impact of different factors on the law of gas kick in deepwater drilling in 2.2, in most cases, especially when drilling into shallow high-pressure gas reservoirs such as shallow gas, the mud pool increment when the gas reaches the seabed mud line after gas kick in deepwater drilling is much lower than the 1–2 m^3 warning value specified in the marine drilling manual, which is not conducive to timely and accurate detection of gas kick. Taking Fig. 2.1 as an example, after a gas kick occurs, when the gas in the wellbore annulus reaches the vicinity of the seabed mud line, the increment of the mud pool is only 0.56 m^3, which is much lower than the warning value of 1 m^3 specified in the marine drilling manual. Even if the increment of the mud pool reaches the warning value of 1 m^3, it cannot be immediately judged whether overflow has occurred. It is necessary to continue to observe whether the return flow rate of the drilling fluid increases and whether the drilling fluid automatically overflows from the return pipe after stopping the pump Whether the increment of the mud pit continues to increase (up to 2 m^3), whether the drilling pressure has decreased, and whether the hanging weight of the hook has increased. When the mud pit increment reaches 2 m^3, the gas has been transported in the riser for nearly 9 min, rising by 558 m. At this time, a large amount of gas kick gas has accumulated in the riser, wasting valuable well control time and greatly increasing the risk of well control.

Meanwhile, according to the analysis in 2.3, the magnitude of the overflow rate during well killing has a significant impact on the peak casing pressure during well killing. For gas kick formations with high original formation pressure, there is a risk of well killing failure if the overflow rate is 1 m^3. It can be seen that the existing 1 m^3 overflow warning value is not applicable to some high-risk deepwater wells.

Based on the above analysis, the limitation of the mud pool increment method in deepwater drilling is that when gas kick occurs, the mud pool increment response is relatively delayed due to the high static pressure environment in deepwater, and the 1–2 m^3 warning value specified in the offshore drilling manual is too conservative for wells with deeper water depth, which cannot guarantee the safety of well killing; According to current operating procedures, even if the mud pit increment reaches 1 m^3, it is not possible to accurately determine whether overflow has occurred. The reason for this is that the warning value is an empirical value, which has poor universality and reliability.

Therefore, it is necessary to design specific overflow warning values for regions or specific well locations, fully considering gas migration laws and well killing safety, and improving the accuracy and reliability of overflow monitoring for this warning value. In this way, for deep water wells with higher risks, overflow can be detected in a timely and accurate manner, while for wells with lower risks, the frequency of false alarms can be reduced, reducing monitoring costs.

2.4.2 Back Calculation Method for Overflow Warning Values in Deepwater Drilling Based on Gas Kick

The basic principle of using the mud pit increment method to determine overflow is to measure the increment of drilling fluid volume in the circulating pool through instruments to reflect the amount of bottomhole overflow flow, and to set a certain warning value based on experience to eliminate the influence of various factors. It is believed that overflow occurs when the warning value is exceeded.

This method has the characteristics of simple operation, convenience and practicality, and high universality for conventional drilling. Deepwater drilling places higher demands on well control. Factors such as deepwater environment significantly reduce the time left for well control in the event of gas kick. Therefore, it is necessary to timely and accurately detect the occurrence of overflow or gas kick, while ensuring sufficient time for well control measures such as well killing. From this, it can be seen that the conventional mud pit increment method cannot meet the well control requirements of deepwater drilling. It is necessary to comprehensively consider factors such as the gas migration law and well killing safety after overflow or gas kick.

According to the analysis in 2.2, when gas kick gas migrates to the vicinity of the seabed wellhead in deepwater drilling, the overflow rate often does not reach the warning value of 1 m^3. Therefore, the location of gas kick gas migration can be used as a parameter for designing the overflow rate warning value. At the same time, in order to incorporate factors of well killing safety and achieve overflow warning values that vary from block to block, and even from well to well, the maximum allowable shut in casing pressure is introduced to design the overflow warning value.

According to the calculation method of the maximum allowable shut in casing pressure in 2.3, when drilling construction parameters, formation conditions, and other factors are determined, the maximum allowable shut in casing pressure is a constant value. Once a gas kick occurs, if the well is killed under different overflow conditions, there must be a limit value of overflow flow. When the well is killed beyond this value, the peak shut in casing pressure will exceed the maximum allowable shut in casing pressure, and there is a risk of leaking the formation or damaging the wellhead equipment, even leading to well killing failure.

Therefore, in order to improve the monitoring accuracy and reliability of the overflow warning value after the occurrence of gas kick in deepwater drilling, and to obtain the extreme value of overflow warning value to ensure the safety of well killing, the model in 2.1 is used to simulate the gas kick and well killing process of the target deep water well, control the gas kick within the safe well depth range, and calculate the peak closing casing pressure of the well under different overflow conditions using different drilling fluid densities for well killing, And by comparing with the maximum allowable shut in casing pressure, calculate the overflow warning value of the deep water well in reverse. The specific calculation steps are as follows:

(1) Based on the drilling parameters and formation pressure profile data of deep water wells, determine the depth H_q of high-pressure gas layers and other wells that are prone to gas kick, assuming that gas kick occurs when drilling to the depth of the well. Calculate the maximum allowable shut—in casing pressure P_j under different kill fluid density ρ_j conditions after gas kick occurs in this deep—water well;
(2) Using the method in 2.3.1, establish a wellbore pressure calculation model and well killing model for the deepwater well after gas kick, and simulate the corresponding relationship between the gas arrival position and overflow rate in the wellbore after gas kick (principle as shown in Fig. 2.1). Based on the performance of the well control device, the overflow rate when the gas reaches the depth H_0 (near the recommended subsea wellhead) is selected as the upper limit Q_{max} of the overflow rate warning value based on the gas arrival position;
(3) Simulate well killing for different overflow rates, select a step size Q_U according to actual needs, and calculate the overflow rate $Q_i = (1 + i)Q_0$ (where i takes 0–n) using drilling fluids of different densities ρ_i calculate the peak casing pressure $P_{max,ij}$ for well killing, and stop calculating when $P_{max,ij} > P_j$. Record the overflow flow rate Q_K at this time;
(4) The accurate calculation value of the warning value for deepwater drilling overflow based on gas kick can be obtained as follows:

$$Q_W = \min\{Q_{max}, Q_K\} \tag{2.29}$$

In step (2) above, the relationship between gas arrival position and overflow rate was added. By setting the upper limit of overflow rate warning value, the gas was limited below the controllable well depth, which is beneficial for combining with other deepwater drilling well control devices to minimize well control risks.

This method comprehensively considers the location of gas after gas kick and the risk of well killing. The calculated overflow warning value is no longer an empirical value, which can ensure more accurate and reliable gas kick and overflow monitoring under the conditions determined by existing well control equipment, construction parameters, and formation parameters.

2.4.3 *Improved Calculation Method for Overflow Warning Value Based on Reliable Formation Pressure Prediction*

2.4.2 has established a calculation method for the overflow warning value of deepwater drilling based on gas kick. When the local formation pressure and other parameters are determined, the overflow warning value calculated by this method has high accuracy and can meet the needs of gas kick well control in deepwater drilling. However, there are errors in predicting formation pressure, and the errors increase with the increase of water depth and well depth. According to the calculation steps in 2.4.2, if the predicted formation pressure of the high-pressure gas layer is too low, the calculated overflow warning value will be too low, which will increase the frequency of false alarms and monitoring costs; On the contrary, if the predicted formation pressure of the high-pressure gas layer is too high, the calculated overflow warning value will be too high, which will cause the monitoring of overflow warning value to lag and increase the risk of well control. Therefore, it is necessary to introduce a reliable formation pressure prediction method to improve the calculation method of overflow warning values.

Professor Zhichuan Guan et al., based on the principles of geostatistics and probability statistics, addressed the uncertainty of formation pressure information by using a profile with credibility to quantitatively describe the distribution of formation pressure, transforming the formation pressure profile into a distribution interval with certain probability information. The prediction of this method is more reliable than the previous single curve, as shown in Fig. 2.14.

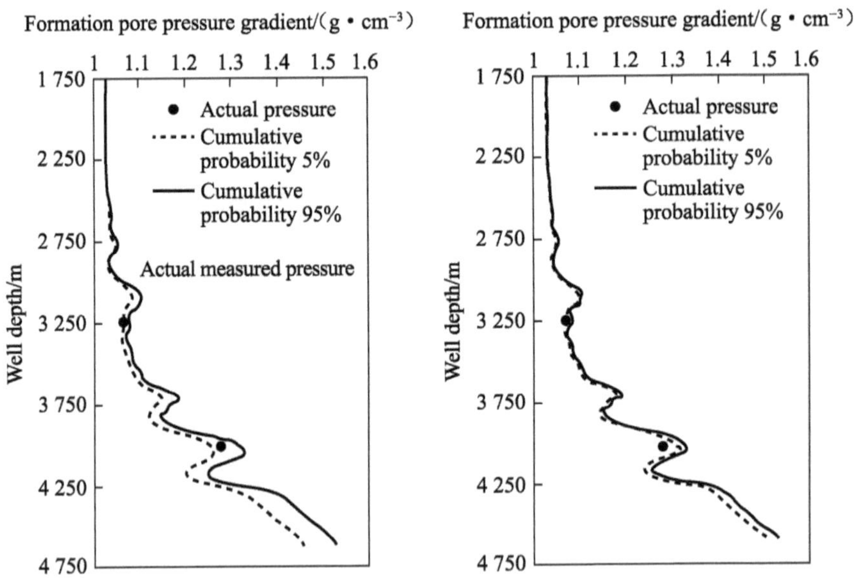

Fig. 2.14 Formation pore pressure profiles with confidence

Figure 2.14 shows the prediction results of formation pore pressure with a reliability of 90% (left figure) and 30% (right figure), respectively. In practical applications, in order to improve prediction accuracy, a higher reliability can be selected for formation pressure prediction.

Assuming that the confidence level J is chosen to predict the formation pressure of deep water wells, the predicted formation pressure is a range of values containing confidence level J information $\left[\mathrm{P_{h1}}, \mathrm{P_{h2}}\right]$, where $\mathrm{P_{h1}}$ represents the lower limit of the predicted formation pressure at well depth h and $\mathrm{P_{h2}}$ represents the upper limit of the predicted formation pressure at well depth h.

The formation pressure used in the reverse calculation method for the overflow warning value in 2.4.2 is no longer a single value, but needs to be calculated by taking points in the interval $\left[P_{\mathrm{h1}}, P_{\mathrm{h2}}\right]$, in order to obtain the upper limit value range of the overflow warning value based on the gas arrival position $\left[\mathrm{Q_{max1}}, \mathrm{Q_{max2}}\right]$ and the overflow value range that reaches the maximum allowable shut-in casing pressure $\left[\mathrm{Q_{K1}}, \mathrm{Q_{K2}}\right]$. At this time, the overflow warning value based on the prediction of formation pressure with confidence is:

$$Q_{\mathrm{W}} = \min\left\{\left[Q_{\mathrm{max1}}, Q_{\mathrm{max2}}\right], \left[Q_{\mathrm{K1}}, Q_{\mathrm{K2}}\right]\right\} \tag{2.30}$$

The Q_{W} in the above equation contains credibility information, which can be selected for design and calculation based on the accuracy of formation pressure prediction, well control device performance, and well control accuracy of the target well.

Therefore, a calculation method for overflow warning value was established, which comprehensive consideration the gas arrival position after gas kick, well killing safety, and uncertainty in predicting formation pressure. This can improve the reliability and accuracy of early monitoring of gas kick in deepwater drilling while reducing costs. At the same time, it can achieve quantitative calculation of overflow warning values that vary from well to well or even from well depth, Solved the potential risk issue of gas kick control in deepwater drilling with a warning value of 1 m^3 in the marine drilling manual.

2.4.4 *Instance Calculation*

Taking a deep water well in the South China Sea as an example, a gas kick occurred when drilling to 3650 m (including a depth of 1310 m). The overflow warning value calculation method established in this chapter was used to design the overflow warning value for the well. The main parameters of the well are as follows:

The water depth is 1310 m, the cemented section is 1490 m, the open hole section is 850 m, the surface temperature of seawater is 15 °C, the outer diameter of the riser is 533 mm, the circulating flow rate is 30 L/s, the drilling fluid density is 1.14 g/ cm^3, the drilling fluid viscosity is 0.027 Pa s, the thermal conductivity coefficient of the drilling fluid is 1.73 W/(m ·° C), the viscosity coefficient is 0.27, the fluidity index is 0.4, the formation porosity is 10%, the formation permeability is 50 md, the thermal conductivity coefficient of the formation is 2.25 W/(m ·° C), and the original formation pressure is 43 MPa.

According to Eq. (2.28), the relationship between the maximum allowable shut in casing pressure and the density of the kill fluid when the well is drilled to 3650 m is shown in Fig. 2.15. According to step 2 of calculation in 2.4.2, calculate the relationship between the position of gas arrival in the wellbore annulus after gas kick and the overflow flow rate, as shown in Fig. 2.16.

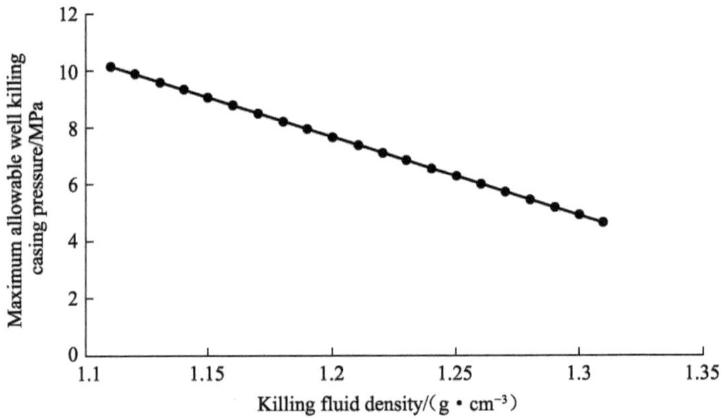

Fig. 2.15 Relation curve of well killing casing pressure and killing fluid density

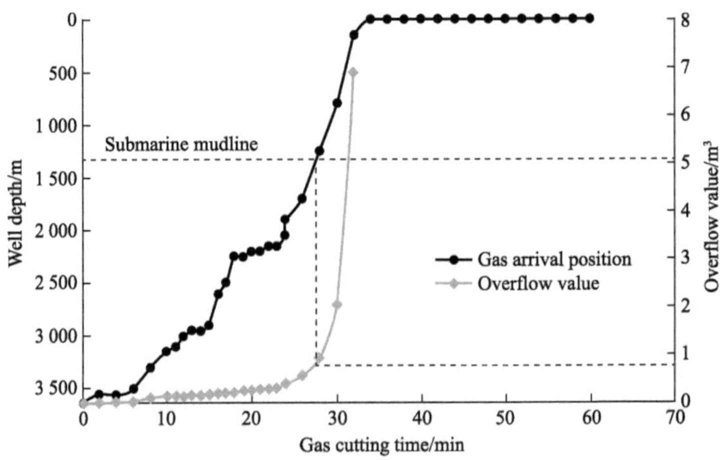

Fig. 2.16 Versus time curve of gas position and overflows

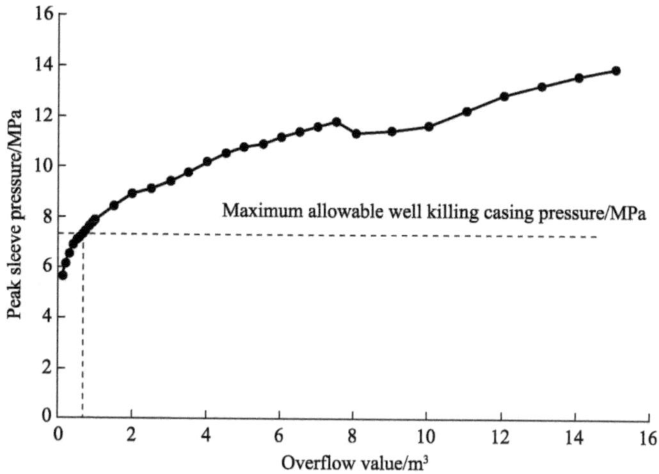

Fig. 2.17 Relation curve of peak casing pressure and overflows

From Fig. 2.16, it can be seen that when the gas kick gas migrates to the seabed wellhead, the overflow flow rate is 0.80 m^3. At this time, the upper limit of the overflow warning value Q_{max} based on the gas arrival position is 0.80 m^3.

According to the drilling design requirements, a well killing fluid with a density of 1.215 g/cm^3 is used for well killing at this time. As shown in Fig. 2.15, the maximum allowable shut in casing pressure at this time is 7.3 MPa. By simulating the well killing process under different overflow conditions, the relationship curve between the peak casing pressure and the overflow rate was obtained, as shown in Fig. 2.17.

As shown in Fig. 2.17, the maximum overflow flow rate $Q_K = 0.68m^3$ is required to ensure that the peak casing pressure does not exceed the maximum allowable shut in casing pressure during well killing.

The accurate calculation value of the overflow warning value for deepwater drilling based on gas kick is:

$$Q_W = \min\{Q_{max}, Q_K\} = \min\{0.80, 0.68\} = 0.68\,m^3$$

It can be seen that if overflow monitoring is carried out according to the conventional mud pit increment method, when the mud pit increment reaches 1 m^3, the gas in the wellbore annulus has already entered the riser section. At this time, if well killing is carried out, the peak casing pressure will inevitably exceed the maximum allowable shut in casing pressure, which poses a risk of damaging the casing, blowout preventer, and lost circulation formation, leading to well killing failure.

If considering the error in predicting the formation pressure of the well based on this, introducing the prediction of formation pressure with confidence, according to the improved calculation method in 4.3, selecting 80% confidence for calculation, the prediction interval of formation pressure at the bottom of the well can be obtained as [41.2, 44.5] MPa. At this point, the upper limit value range Q_{max} of the overflow warning value based on the gas arrival position is [0.68,0.91] m^3, and the overflow value range Q_K of the maximum allowable shut in casing pressure is [0.52, 5.8] m^3. Therefore, the overflow warning value $Q_W = \min\{[0.68, 0.91], [0.52, 5.8]\} = 0.52\,m^3$ based on the prediction of formation pressure with confidence.

For this deep water well, if an overflow warning value of 0.52 m^3 is used to monitor gas kick, the introduction of reliable formation pressure prediction minimizes monitoring errors caused by inaccurate formation pressure prediction, and fully considers the gas arrival position and well killing risk after gas kick, greatly improving the accuracy and reliability of using overflow warning values for early monitoring of gas kick in deep water drilling.

It is worth noting that the overflow warning value designed using the method established in 2.4 has high accuracy, while for semi submersible operating platforms commonly used in deepwater drilling, the liquid level of the mud pit fluctuates greatly, which is not conducive to accurately detecting the increment of the mud pit. To meet the measurement accuracy, it is recommended to use inlet and outlet flow meters with higher accuracy and stability to monitor the increment of the mud pit.

2.5 Early Monitoring and Risk Assessment Methods for Gas Intrusion in Risers

Previously, by analyzing the potential risks of overflow warning values in existing mud pool liquid level monitoring methods, a quantitative design method for overflow warning values was established that comprehensively considers gas arrival location, well killing safety, and formation pressure prediction errors. This method can calculate the overflow warning values that meet safety requirements in real time during the drilling process, achieving the goal of accurately detecting gas kick and controlling well control risks. Although this method quantitatively calculates the warning value of overflow flow, greatly improving the accuracy and reliability of monitoring gas kick, according to the analysis in 2.2, for the drilling section (700–2800 m) with a water barrier installed after surface casing cementing, if drilling into shallow high-pressure gas reservoirs such as shallow gas, the warning value calculated using the method established in 2.4 is too small (the smaller the well section below the mud line, the smaller the calculated warning value), Not conducive to effective monitoring of overflow monitoring equipment; At the same time, it is not possible to quantitatively evaluate the degree of gas kick when it is discovered, and thus cannot provide specific risk indications for subsequent well killing operations. Therefore, on the basis of quantitative calculation of overflow warning values, it is necessary to seek new methods for early monitoring and risk assessment of gas kick in deepwater drilling to meet the needs of well control for gas kick in deepwater drilling.

In recent years, many scholars have conducted research on the early detection of gas kick in deepwater drilling, such as flow differential overflow detection method and acoustic gas kick monitoring method while drilling. The acoustic gas kick detection method utilizes the significant difference in wave velocity of sound waves in the gas–liquid phase for early monitoring of gas kick. Although this method can achieve early monitoring of gas kick, the propagation of sound waves through the drill pipe and the fluid in the annulus, and factors such as external noise caused by the long path, have a significant impact on the monitoring results, resulting in significant errors and instability, and it is difficult to analyze the degree of gas kick at the bottom of the well. In order to achieve more accurate early monitoring of gas kick, domestic and foreign scholars have gradually shifted their research focus to early monitoring methods for gas kick at the riser, mainly including gas risk control technology based on MPD technology, overflow monitoring method based on drilling fluid density measurement at the riser, and gas kick monitoring method based on Doppler principle. Among them, the gas risk control technology based on MPD technology in the riser utilizes the pressure control advantages of MPD technology itself and fluid detection devices to process the gas inside the riser to reduce the risk of well blowout; The overflow monitoring method based on the density measurement of drilling fluid at the riser is determined by measuring the changes in density using a fluid density detection device installed at the bottom of the riser. By analyzing the changes in fluid density in the annulus of the riser, the occurrence of overflow or gas kick is determined. This method is proposed by E.D. Toskey et al. at the 2015 OTC meeting to promote further development of early monitoring technology for overflow at the riser. The method based on measuring density changes cannot accurately reflect the fluid properties and gas content inside the riser. Although this method can qualitatively detect the occurrence of overflow or gas kick, further research and development are needed on how to accurately determine the gas content in the cross-section; The ultrasonic Doppler gas intrusion monitoring method proposed by domestic scholars for riser is based on the frequency difference generated by the reflection of ultrasonic waves by solid particles or bubbles in the fluid in the annulus of the riser to measure the flow velocity of different phase fluids, and thus evaluate the gas content in the cross-section; The method proposed by the author for monitoring gas kick through ultrasonic transmission in the riser utilizes the propagation and attenuation characteristics of ultrasonic waves in multiphase flow in the riser annulus to quantitatively monitor the gas content in the cross-section. At the same time, for deepwater well control, timely detection of gas kick is no longer sufficient to meet the needs of fine well control. It is necessary to combine the monitoring data of gas kick from the riser and the multiphase flow model of deepwater drilling wellbore to simulate the situation of bottomhole gas kick in real time, providing technical support for reasonable and accurate control of gas kick.

2.5.1 *Feasibility Analysis of Monitoring Gas Kick at Deep-Water Riser*

Firstly, analyze the feasibility of early monitoring of gas kick at deep-water risers.

In 2.2, the gas kick law of a deep water well in the South China Sea was calculated and analyzed, as shown in Fig. 2.1. When the gas kick gas was transported to the seabed wellhead, the overflow rate had not yet reached the warning value of 1 m^3. At the same time, by changing different conditions for simulation calculations, it was found that in most cases, the overflow rate when the gas kick reached the seabed wellhead did not exceed this value, and the deeper the water depth, the shallower the depth below the mud line, the lower the bottomhole pressure difference, and the lower the formation permeability, The smaller the overflow flow rate of gas kick when it reaches the seabed wellhead. Therefore, if gas kick is effectively monitored at the bottom of the riser, it can be detected earlier than conventional methods. This is similar to the purpose of the method in 2.4, which can control the gas kick in a controllable well section below the riser, which is beneficial for subsequent well control safety.

Meanwhile, as shown in Fig. 2.1, it takes nearly 30 min for gas to migrate from the bottom of the riser to the wellhead above water. The time required for gas to enter the bottom of the riser and reach the 1 m^3 warning value is nearly 3 min, and the time required to reach the 2 m^3 warning value is 8 min. Therefore, if gas can be effectively detected at the bottom of the riser, it can earn enough time to take well control measures such as well killing compared to conventional methods.

Calculate the variation of gas content at the bottom section of the riser over time after gas kick, as shown in Fig. 2.18.

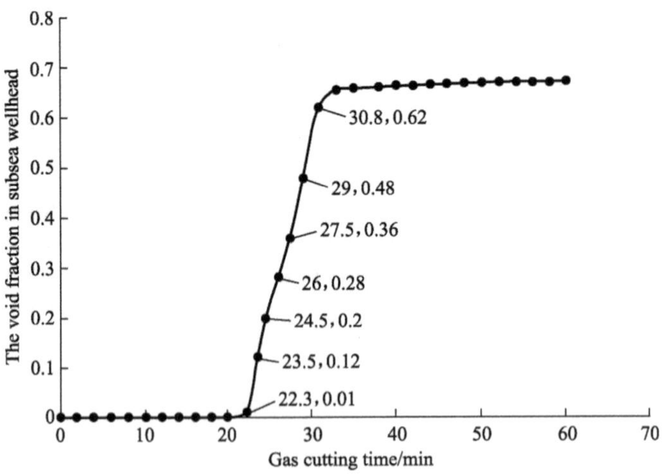

Fig. 2.18 The void fraction changes over time in subsea wellhead

From Fig. 2.18, it can be seen that after the gas reaches the bottom of the riser (22 min), the gas content in the cross-section reaches its peak within 8 min, and it only takes 2 min for the gas content to increase from 0 to 23%. For different bottomhole gas influx rates, the variation of cross-sectional gas content over time is different. Therefore, monitoring the variation of cross-sectional gas content at the bottom of the riser over time can reveal the situation of bottomhole gas influx.

Based on the above analysis, for example deep water wells, monitoring the gas content at the bottom of the riser can accurately detect gas kick several minutes earlier than conventional methods. However, for shallow drilling with larger water depths and the installation of a riser after surface cementing, this method can serve as a powerful supplement to the method in 2.4, and is expected to quantitatively evaluate the magnitude of gas kick at the bottom of the well based on the variation of gas content over time. This method is feasible in both theory and technology.

2.5.2 *Ultrasonic Monitoring Technology for Gas Kick in Riser Pipes*

2.5.2.1 Ultrasonic Doppler Monitoring Method

The riser in deepwater drilling is always under pressure and is in a complex mechanical state due to factors such as undercurrent and drilling vessel drift. Therefore, monitoring gas kick here requires ensuring the integrity of the riser. Therefore, some scholars have proposed using a Doppler ultrasound probe on the outside of the riser to detect gas kick, and the effectiveness of this method for early detection of gas kick has been verified through indoor and on-site experiments. At the same time, the response of ultrasonic frequency shift to gas content has been obtained. By analyzing its measurement principle, it can be concluded that the Doppler ultrasound measurement method is mainly based on the frequency difference generated by the reflection of solid particles or bubbles in the fluid to calculate the flow velocity. According to the theory of multiphase flow, the upward velocity of bubbles in the wellbore annulus is greater than the fluid velocity, while the velocity of solid particles such as rock cuttings is lower than the fluid velocity. The solid phase of the drilling fluid itself is equal to the fluid velocity. Therefore, the results measured by Doppler ultrasound method comprehensively reflect various phase velocities, and the gas content can be reflected by the changes in the measured flow velocity. The problems that still need to be solved by this method are as follows:

(1) Extraction of bubble rising speed signal: How to effectively distinguish the ultrasonic signal characteristics reflected by different phase fluids and accurately measure the actual rising speed of bubbles.
(2) Simulation of multiphase flow field in the riser annulus: The ultrasonic Doppler measurement method adopts a combination of oblique probe emission and reception, and its measurement information window is located at the edge of

the wellbore annulus flow field. How to use the measured data to reflect the flow velocity of the entire annulus section still requires in-depth analysis of the flow field distribution.

2.5.2.2 Ultrasonic Transmission Monitoring Method

The author proposed a method for quantitatively monitoring the gas content in the cross-section using ultrasonic transmission and the propagation and attenuation characteristics of ultrasonic waves in multiphase fluids in the riser annulus, and conducted preliminary experimental and numerical simulation studies. The experimental setup is shown in Figs. 2.19 and 2.20.

As shown in Fig. 2.19, this experimental setup is designed in a ratio of 1:2 to the actual size, which can simulate multiphase flow in the annular space of a riser under different cross-sectional gas content conditions. The response of sound waves to gas content is measured in real time using an ultrasonic probe to analyze the quantitative characterization relationship between ultrasonic signal characteristic parameters and wellbore cross-sectional gas content. The wellbore is designed with a combination of upper stainless steel and lower organic glass. The upper stainless steel wellbore can be used to simulate the riser, while the lower organic glass is convenient for observing the movement of bubbles in the wellbore. In the experiment, the activation of the gas storage tank can be determined based on the actual gas flow rate: when the gas flow rate is small, close valve 2 and open valve 1. The gas generated by the air compressor enters the simulated wellbore through the pressure regulating valve, flow meter, throttle valve, and valve 1, and forms a uniform bubbly flow through the foam sieve plate at the bottom. The size of the bubbles in the simulated wellbore can be changed by changing the number and size of the holes on the foam sieve plate;

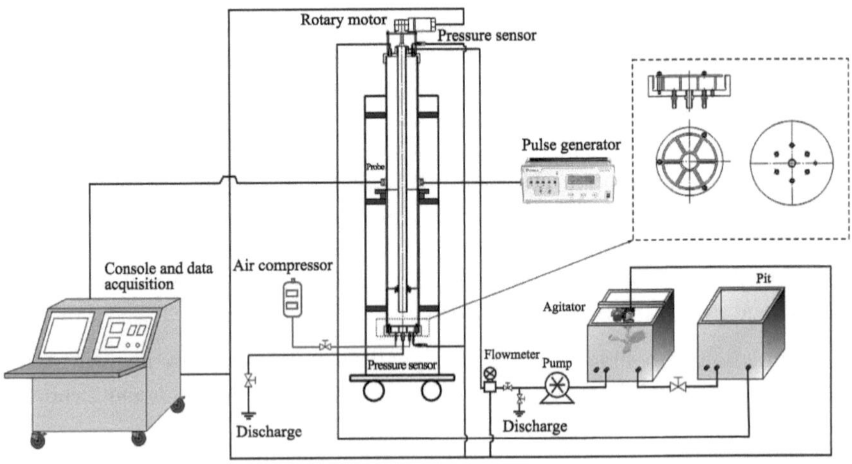

Fig. 2.19 Simulation device of gas cut monitoring using ultrasound

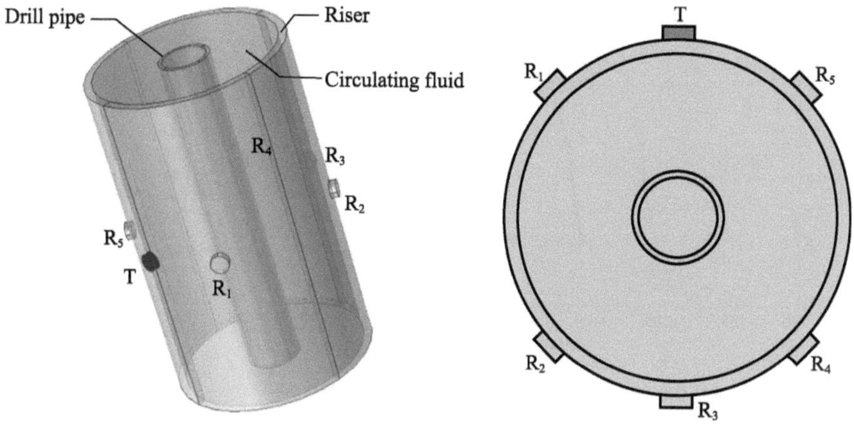

Fig. 2.20 Ultrasonic probe installation position

When the air flow rate is high, close valves 1 and 2, open the air compressor to fill the storage tank (to stabilize the gas), open valve 2, and control the intake volume by adjusting the throttle valve.

As shown in Fig. 2.20, the ultrasonic probe is installed on the wall of the water barrier by adding a delay block. The ultrasonic generator adopts a 900 V high voltage generating device, which is conducive to achieving two-phase flow of ultrasonic penetrating gas at different frequencies, in order to study the effect of frequency on ultrasonic attenuation law and select the optimal frequency.

In the experiment, it is necessary to accurately determine the gas content of the cross-section in order to quantitatively analyze the relationship between ultrasonic signal characteristic parameters and the gas content of the cross-section. At present, the main method for measuring the gas content in the cross-section is the probe method, which has low accuracy and cannot meet the needs of finely adjusting the gas content in this experiment. Therefore, based on the size of the experimental device, Fluent software was used to simulate the gas content of the wellbore section under different intake conditions, in order to establish a quantitative relationship between the gas content of the experimental device section and the gas flow rate.

The distribution of gas content in each section of the wellbore under different gas flow rates is as follows.

Figure 2.21 shows the distribution of gas content in the cross-section of the entire simulated wellbore under different gas flow rates. Due to the low gas content level of the simulated bubbly flow, there is no flow pattern change during the rising process of the bubbles. The gas content in the cross-section is basically linearly distributed along the wellbore, and the change amplitude is small. Taking the section at 1.4 m as an example, the relationship between gas flow rate and section gas content can be obtained, as shown in Fig. 2.23.

According to the results in Fig. 2.22, the empirical formula for fitting the gas content and gas flow rate at a cross-section of 1.4 m is:

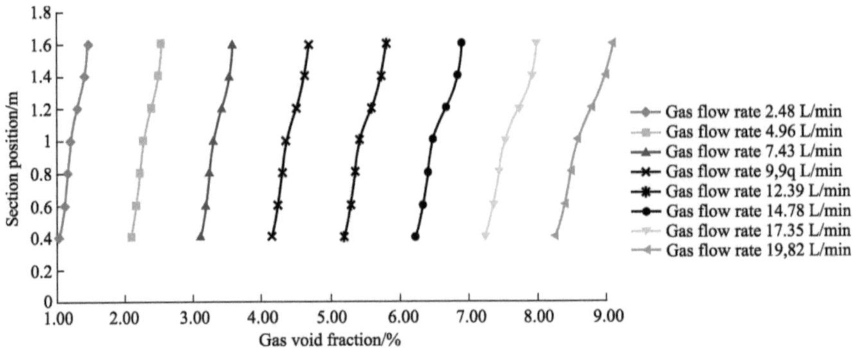

Fig. 2.21 The gas fraction distribution under different gas flow rate

$$\phi(h) = 0.0044Q_g + 0.0032$$

Among them, $\phi(h)$ is the gas content of the cross-section at h; Q_g is the gas flow rate, L/min.

From this, the size of the cross-sectional gas content can be determined in real-time based on the gas flow rate.

Using ultrasonic transmission method, the attenuation characteristics of ultrasonic waves in gas–liquid two-phase flow are used to reflect the cross-sectional gas content, which has the following advantages:

(1) The range of ultrasonic penetration cross-section is larger, and the measured data (such as sound amplitude and speed) can more accurately and comprehensively reflect the gas content of the cross-section;
(2) The reflection effect of bubbles on ultrasound can be ignored, and only the signal received by the receiving probe needs to be processed;

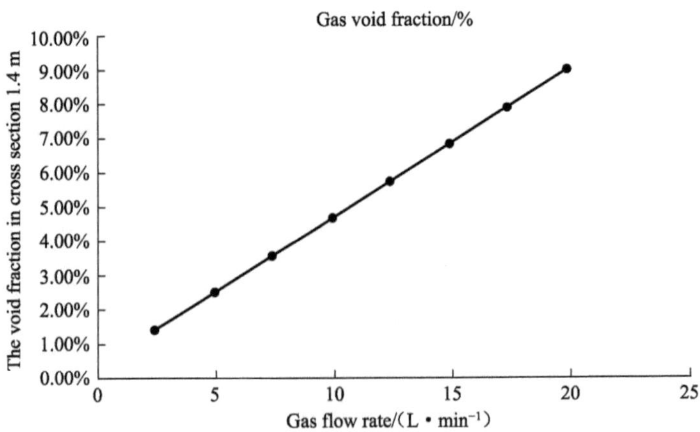

Fig. 2.22 The relationship between gas fraction and gas flow rate in cross section 1.4 m

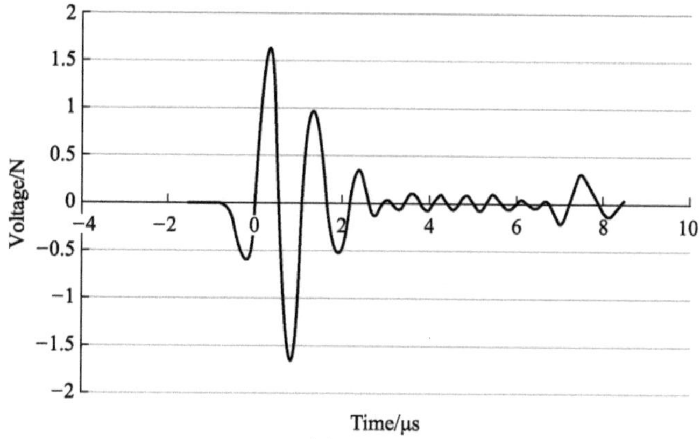

(a) Q_g=0 L/min

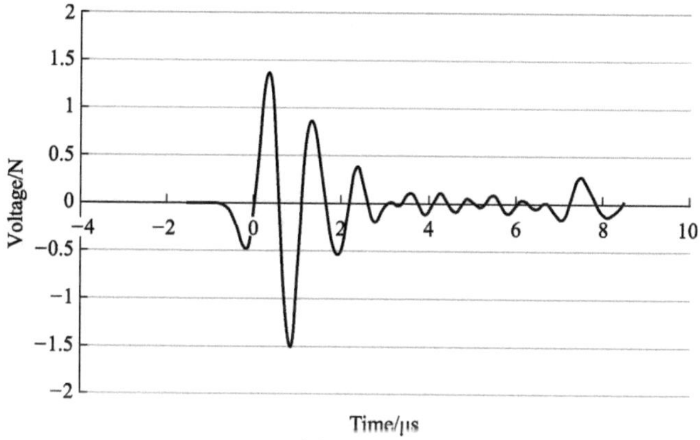

(b) Q_g=2.8 L/min

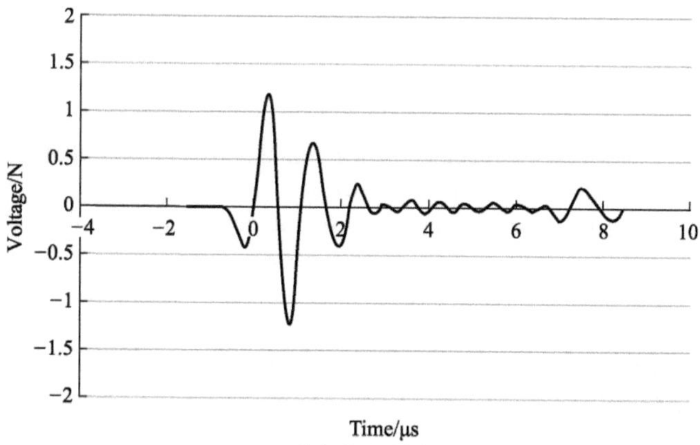

(c) Q_g=7.7 L/min

Fig. 2.23 The ultrasonic receiving signals under different gas rate

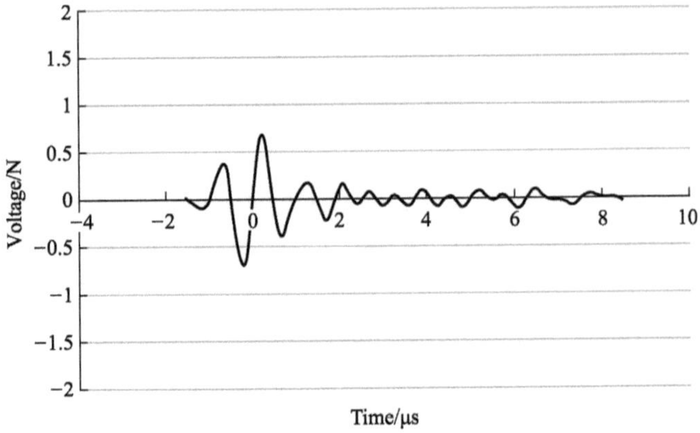

(d) Q_g=12.6 L/min

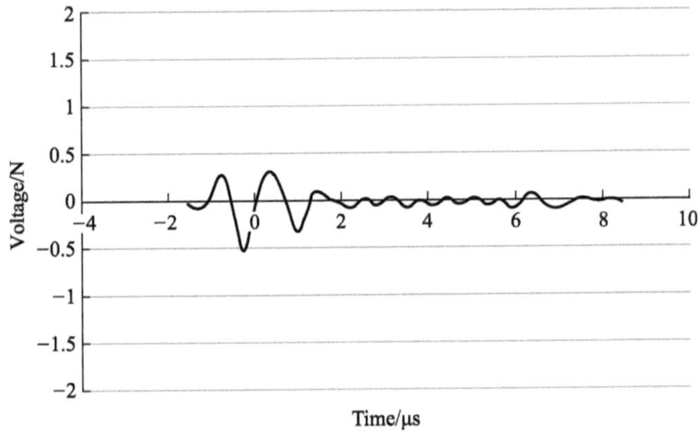

(e) Q_g=18.9 L/min

Fig. 2.23 (continued)

(3) Multiple experiments can be conducted to quantitatively determine the correspondence between the received probe signal and the gas content of the cross-section under certain conditions (drilling fluid performance, rock cuttings concentration, displacement, etc.), in order to achieve a quantitative description of bottomhole gas kick.

2.5.3 Indoor Experiments and Numerical Simulation Analysis

Based on the ultrasonic transmission monitoring method established in 2.5.2.2, experimental and numerical simulation methods were used to test and analyze the effectiveness of the ultrasonic transmission monitoring method, in order to demonstrate the feasibility of using the ultrasonic transmission monitoring method to accurately determine the cross-sectional gas content in the riser.

2.5.3.1 Experimental Research

Select the cross-section at 1.4 m of the simulated riser in Fig. 2.19 as the experimental monitoring point, and use two ultrasonic single crystal straight probes with a frequency of 0.1 MHz as the transmitting and receiving probes, respectively. The probes are installed on the outer wall of the riser through delay blocks and coupling agents, as shown in Fig. 2.20. Conduct experiments by changing the intake volume to 0 L/min, 2.8 L/min, 7.7 L/min, 12.6 L/min, and 18.9 L/min (with simulated cross-sectional gas content of 0%, 1.55%, 3.71%, 5.87%, and 8.65%, respectively), and record the acoustic signal data at the receiving end, as shown in Fig. 2.23.

From Fig. 2.23, it can be seen that under different gas flow rates (cross-sectional gas content), there is a significant change in the amplitude of the ultrasonic received signal, which is a sensitive parameter. Therefore, the degree of ultrasonic attenuation can be used as one of the quantitative indicators for evaluating the gas content of the riser annulus section. In future experimental research, it is necessary to further simulate the response law of ultrasonic reception signals to cross-sectional gas content under different drilling fluid densities, displacements, and other conditions. Based on a large amount of experimental data, a calculation model that characterizes the degree of ultrasonic attenuation should be established to establish a quantitative characterization relationship between ultrasonic signal characteristic parameters and cross-sectional gas content. By the time of manuscript submission, the experimental research in this section has been relatively in-depth and comprehensive, and the influence laws of different factors on sound wave propagation have been obtained. Specific achievements can be referenced in the relevant journal papers published by the author.

2.5.3.2 Numerical Simulation Research

The physical field of the system composed of the riser, drill pipe, fluid, and bubbles in the experiment is relatively complex. Ultrasonic waves propagate in the three physical fields of solid, liquid, and gas, requiring coupling calculations of multiple physical fields. The multi physical field numerical simulation software COMSOL Multiphysics can be used to simulate the system. By analyzing the multiphase flow field and ultrasonic propagation law in the wellbore annulus, the model is simplified as follows:

(1) In order to reduce computational workload, the size of the model was reduced by 10 times compared to the experimental setup for modeling. According to the principle of similarity, the ultrasonic frequency should be increased to 10 times the experimental frequency.

(2) Due to the fact that the propagation speed of ultrasound in multiphase fluids in the wellbore annulus is much faster than the migration speed of fluids and bubbles, when simulating the attenuation law of ultrasound (without considering the Doppler effect caused by bubble velocity), it can be considered that both the fluid and bubbles are in a stationary state.
(3) Due to the fact that the distribution of bubbles in the cross-section of the wellbore is random in practical situations, uniform distribution of bubbles or the definition of a random function can be used to simulate different gas content situations in the cross-section.
(4) The riser (solid) satisfies the mechanical properties of solids and belongs to linear elastic materials.

Establish a physical model of ultrasonic propagation for gas–liquid two-phase flow in a riser based on the above simplification principles. Taking the simulation of a gas content of 1.55% as an example, the physical model of ultrasonic propagation is shown in Fig. 2.24.

In Fig. 2.24, there are 20 uniformly distributed bubbles with a diameter of 0.6 mm (simulating a real bubble size of 6 mm) in the annulus of the riser, and the gas phase content in the annulus is 1.55%. Add a 1 MHz Gaussian pulse displacement excitation to the upper boundary of the delay block to generate 1 MHz ultrasonic waves, and the lower boundary probe is used to receive acoustic signals.

The solid mechanics model was used for the delay block, riser, and drill pipe, and the pressure acoustic model was used for the multiphase fluid in the annulus. The transient research module was used to study the propagation law of ultrasonic waves. The sound pressure distribution in the multiphase fluid in the riser annulus at each moment is shown in Fig. 2.25.

The signal received by the lower boundary probe is shown in Fig. 2.26.

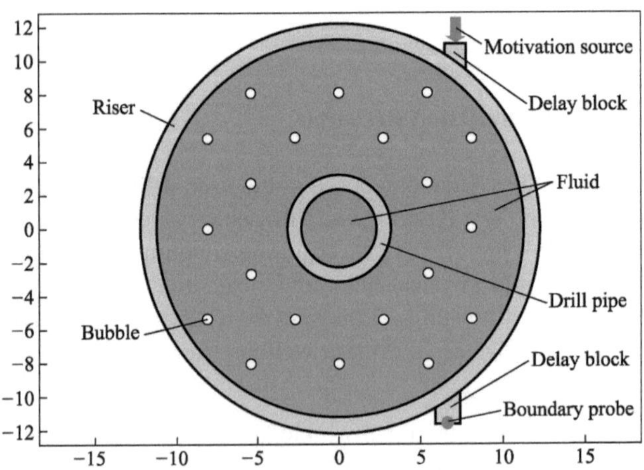

Fig. 2.24 The physical model of ultrasonic wave propagation in gas–liquid two-phase flow of riser

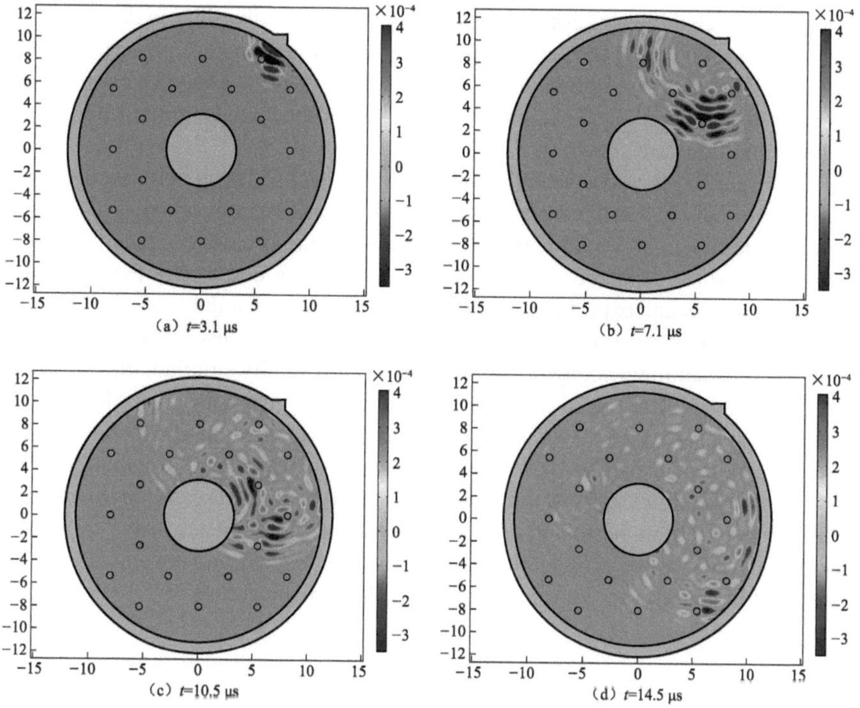

Fig. 2.25 The ultrasonic sound pressure distribution in gas–liquid two-phase flow of riser at different moment

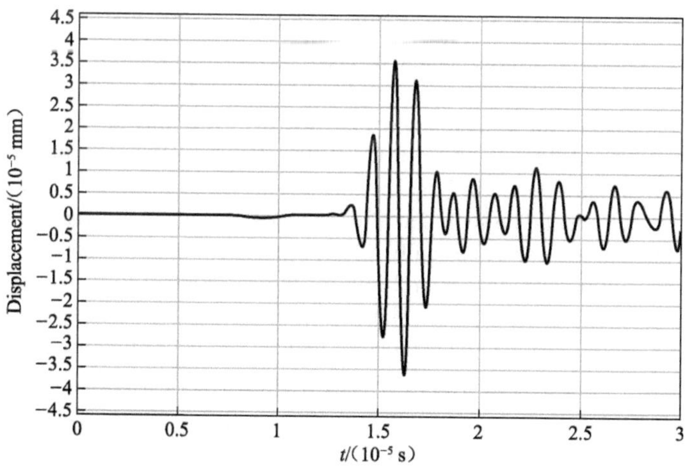

Fig. 2.26 The probe signal when gas fraction is 1.55%

Keeping the bubble size constant, changing the bubble distribution density in the riser annulus, simulating the propagation law of ultrasound in multiphase flow in the riser annulus under conditions of 0%, 3.71%, 5.87%, and 8.65% gas content, and recording the received probe signal. The received probe signals under different gas content conditions are shown in Fig. 2.27.

From Fig. 2.27, it can be seen that there is a strong correlation between the attenuation degree of the received probe signal and the gas content under different gas content conditions. At the same time, the time when the peak of the received signal appears also shifts with the increase of gas content. By comparing the numerical simulation results and experimental results, it can be seen that the numerical simulation results are basically consistent with the experimental results, indicating that the model established in this chapter can be used to numerically simulate the ultrasonic propagation characteristics in gas–liquid two-phase flow in the riser, in order to study the ultrasonic propagation laws in gas–liquid two-phase flow in the riser under more different conditions (such as different bubble sizes, densities, and distributions). By the time of manuscript submission, the numerical simulation research in this section had been relatively in-depth and comprehensive, and the influence laws of different factors on sound wave propagation were obtained. Specific achievements can be referenced in the relevant journal papers published by the author.

2.5.4 Quantitative Description Method of Wellbore Gas Kick Degree Based on Monitoring Data of Riser

If gas kick is detected at the riser using ultrasound, it is necessary to determine whether appropriate measures need to be taken for well control based on the gas kick situation, and the judgment is based on gas kick time and degree. The degree of gas kick at the bottom of the well can be calculated by combining gas kick monitoring data at the riser and the two-phase flow model of air and liquid in the deep water drilling wellbore introduced in 2.1.

From Fig. 2.18, it can be seen that after gas kick, the change in gas content at the bottom of the riser section is basically linear, especially when the gas content is low (<20%), and it only takes less than 2 min for the gas content to increase from 0 to 20%, indicating strong timeliness. Therefore, the gas content and its variation rate at the section of the riser can be used as an indicator to determine the degree of gas kick in the wellbore. The specific steps are as follows:

(1) Establish a wellbore gas–liquid two-phase flow calculation model based on the wellbore structure, drilling fluid performance, construction parameters, and formation parameters of the target well;
(2) As the depth of the well increases, the error in predicting formation pressure also increases. If the single value of formation pressure is directly used as the basic parameter to calculate the gas kick process, it is difficult to accurately reflect the situation of bottomhole gas kick. Therefore, the confidence based formation

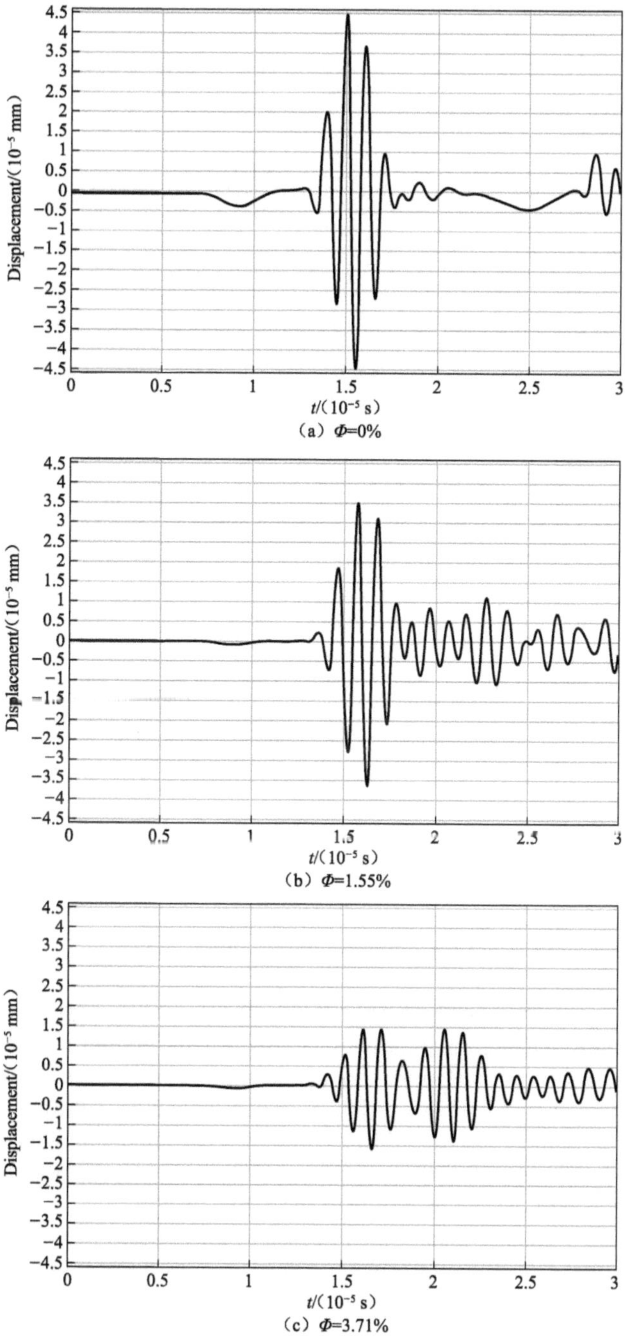

Fig. 2.27 The probe signal with different gas fraction

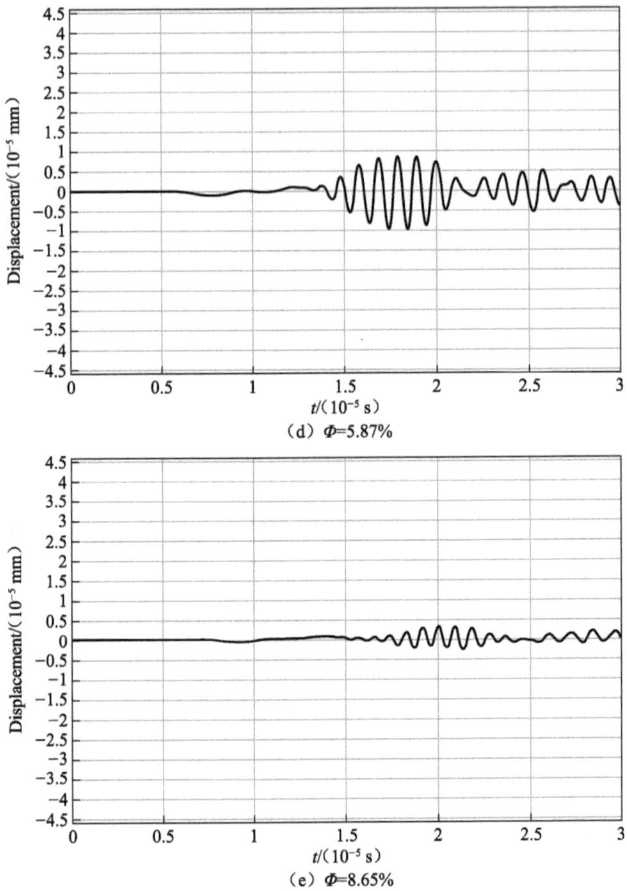

Fig. 2.27 (continued)

pressure prediction method described in the previous chapter is introduced. Assuming that the original formation pressure at the bottom of the target well is P_0, the confidence based formation pressure range at the bottom of the target well is determined by calculation as $[P_1, P_N]$, corresponding to the confidence level of J. The wellbore gas–liquid two-phase flow calculation model established in step 1 is used to simulate the gas kick process under different original formation pressure conditions at the bottom of the well;

(3) Obtain the variation of gas content at the bottom section of the riser with time under a series of different original formation pressures $[P_1, P_N]$ (the trend is shown in Fig. 2.18). Select the time point at which the gas content starts non-zero $[T_{01}, T_{0N}]$ and the time point at which the gas content reaches EgM (selected from 0–20% according to actual needs) $[T_{t1}, T_{tN}]$ Calculate the variation rate of gas content (0 – EgM) at the riser over time:

$$K_i = \frac{E_{gM}}{T_{ti} - T_{0i}} \tag{2.31}$$

Then T_{0i} and K_i are characteristic parameters that reflect the degree of gas kick at the bottom of the wellbore by the cross-sectional gas content at the riser.

(4) Assuming that the acoustic monitoring method introduced in 2.5.2.2 is used to monitor Δt The change rate of the gas content at the bottom section of the riser during time t is K_c, and the time when bubbles are detected is T_c. If:

$$\left|K_c - K_j\right| < \varepsilon \tag{2.32}$$

It is believed that the original formation pressure P_j corresponding to K_j at the bottom of the well is the true formation pressure. among ε Determine the value based on the actual situation.

(5) In step 2, select the wellbore gas kick simulation results corresponding to P_j, determine the time it takes for the bubbles to reach the bottom of the riser, that is, when the signal is detected, the gas kick has already occurred at T_{0j}, and the time when the gas kick occurs is $T_c - T_{0j}$. At the same time, calculate the time T_{lim} required for the gas to reach the wellhead and the total overflow flow rate Q_{ZY}, so that the size of the total overflow flow rate after the gas kick can be determined in real time based on the cross-sectional gas content at the riser. In addition, the gas distribution in the wellbore annulus at each moment after gas kick can be calculated, that is, the cross-sectional gas content $E_g(H, t)$ at different depths of the well.

Thus, based on the acoustic monitoring data at the bottom of the riser, the timing of gas kick $T_c - T_{0j}$ the remaining time T_{lim} for gas to reach the wellhead, the total overflow flow rate Q_{ZY} at any time, and the cross-sectional gas content $E_g(H, t)$ at different depths of the well at any time can be accurately determined, providing a real-time quantitative description of the gas kick situation in the wellbore.

2.5.5 *Instance Calculation*

Taking a deep water well in the South China Sea as an example, calculate the position of gas arrival in the wellbore annulus and the corresponding mud pool increment after a gas kick occurs. The relevant parameters are as follows:

The water depth is 1524 m, the well depth is 5030 m, the cemented section is 3153 m, the surface temperature of seawater is 15 °C, the outer diameter of the riser is 508 mm, the circulating flow rate is 28 L/s, the drilling fluid density is 1.29 g/cm^3, the thermal conductivity of the drilling fluid is 1.73 W/(m·°C), the viscosity coefficient is 0.27, the fluidity index is 0.4, the formation porosity is 10%, the permeability is 50 md, the thermal conductivity of the formation is 2.25 W/(m·°C), and the original formation pressure is 70 MPa.

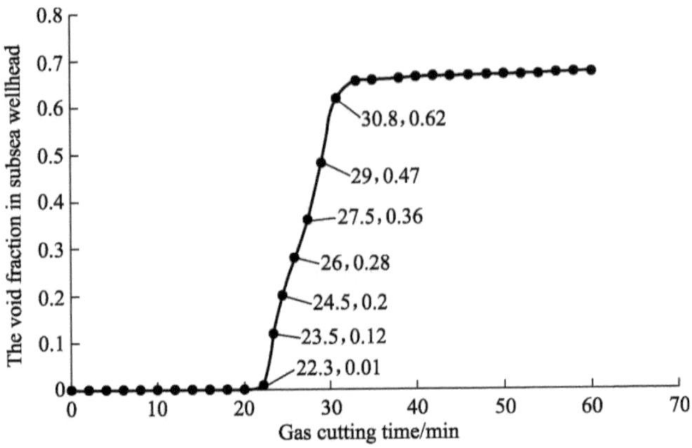

Fig. 2.28 Change rate of void fraction in riser over time

Assuming that the moment of gas kick occurs at 0 o'clock, the ultrasonic probe at the bottom of the riser detects the moment of gas kick as T_c, and the rate of change in the gas content of the cross-section at the riser is $K_c = 0.09$; Based on neighboring well logging and seismic data, a 90% confidence level is selected to predict the bottomhole pressure. The range of bottomhole pressure values is [67.5,72.9], and E_{gM} is taken as 10%. The change rate of gas content at the section of the riser corresponding to the bottomhole pressure value within the range is calculated, which is $K_i \in [0.082, 0.142]$ According to Eq. (2.31), $K_j = K_c = 0.09$ is determined, and the corresponding wellbore pressure value is 68.2 MPa. This value is considered to be the true value of the wellbore pressure. The variation rate of the gas content at the section of the riser with time is shown in Fig. 2.28.

When gas kick was detected at the riser, $T_{0j} = T_c = 22.3$ min had already occurred. At the same time, by simulating the relationship between gas arrival position and time (as shown in Fig. 2.29), the remaining time for gas to reach the wellhead $T_{lim} = 7.6$ min is obtained.

The variation of total overflow over time is shown in Fig. 2.30.

When gas kick is detected at the riser, the distribution of gas content in the cross-section of the wellbore at $t = T_{oj} = 22.3$min is shown in Fig. 2.31.

Therefore, necessary well control measures can be taken based on the relevant data of T_{lim}, Figs. 2.30 and 2.31.

2.6 Summary

(1) By analyzing the overflow rate of gas reaching the seabed wellhead after gas kick under different water depths, depths below the mud line, bottomhole pressure difference, and formation permeability conditions, as well as the gas arrival

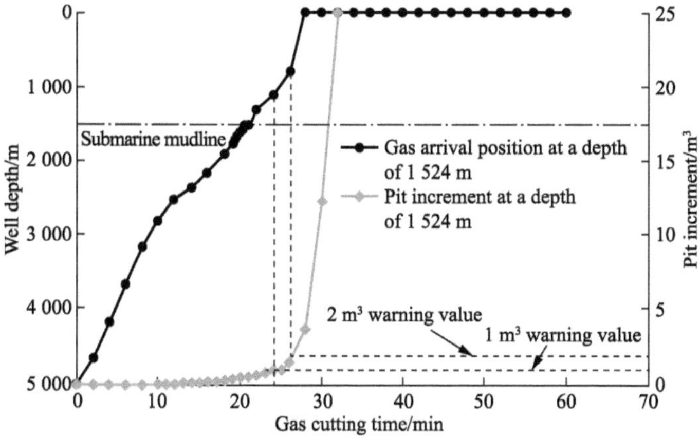

Fig. 2.29 The gas position in wellbore and increment of mud pool after gas cut

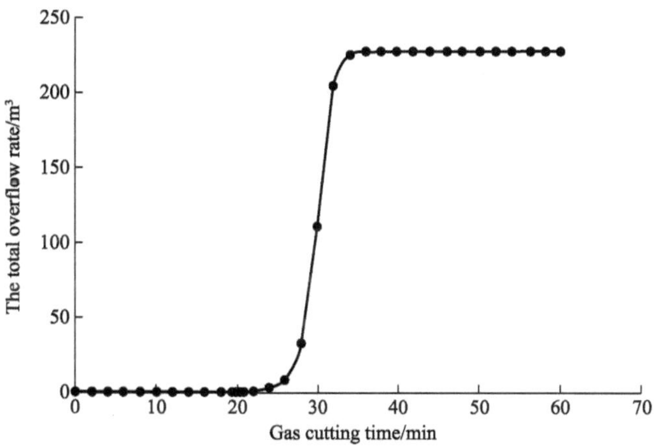

Fig. 2.30 The total overflow rate changes over time after gas cut

positions at overflow rates of 1 m^3 and 2 m^3, the following conclusions can be drawn:

(a) For deep water drilling (with a depth greater than 800 m), the high static pressure environment in deep water can result in a small increase in mud pool increment when gas reaches the seabed wellhead, and in most cases, it cannot reach the 1 m^3 warning value specified in the offshore drilling manual.

(b) The smaller the depth below the mud line, the more unfavorable the existing mud pool liquid level monitoring method is for monitoring gas kick in deepwater drilling. If shallow gas reservoirs (buried within 1500 m below the mud line) are encountered during deepwater drilling, it is difficult to

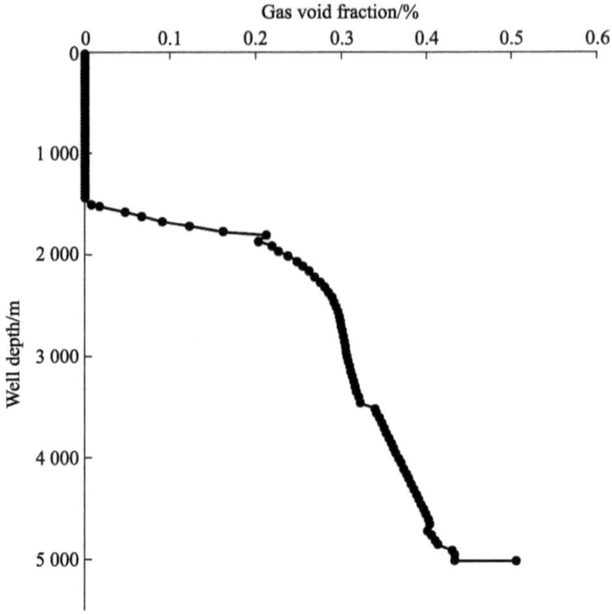

Fig. 2.31 The gas void fraction distribution in wellbore when gas reaches the subsea wellhead

detect gas kick in a timely manner using the existing mud pool liquid level monitoring method.

(c) For deepwater drilling, the magnitude of the bottomhole pressure difference directly affects the overflow rate of gas when it reaches the seabed wellhead and the distance of gas migration in the riser when the mud pool increment reaches 1 m^3 and 2 m^3. Therefore, accurate prediction of formation pressure is crucial for controlling gas kick in deepwater drilling.

(d) The permeability of the formation has a significant impact on the gas kick in deepwater drilling. The smaller the permeability of the formation, the more unfavorable it is for the mud pool liquid level monitoring method to monitor the gas kick in deepwater drilling.

(2) By analyzing the risk factors and influencing laws of gas kick and well killing, combined with the analysis of gas kick and overflow laws, it is found that the overflow rate during well killing has a significant impact on well killing risk. However, the existing 1 m^3 overflow rate warning value is an empirical fixed value, and its safety value is uncertain. For wells with deep water depth, shallow depth below the mud line, and high original formation pressure, it may not meet the requirements for well killing safety, Seeking an accurate method for calculating this value; The prediction of formation pressure has errors, and its value has a significant impact on the peak casing pressure (well killing risk). Therefore, when calculating the overflow warning value, the influence of formation pressure prediction errors should be considered.

(3) By introducing factors such as the maximum allowable shut in casing pressure and the location of annular gas arrival, a reverse calculation method for overflow warning values based on gas kick has been established. This method can accurately calculate the overflow warning values for specific well locations and depths, ensuring more accurate and reliable gas kick and overflow monitoring under the conditions determined by existing well control equipment, construction parameters, and geological parameters. At the same time, the introduction of a reliable formation pressure prediction method improves the calculation of overflow warning values, eliminating monitoring errors caused by inaccurate formation pressure prediction to the greatest extent possible. This is beneficial for selecting different levels of reliability to design overflow warning values based on actual site conditions. To ensure sufficient monitoring accuracy, it is recommended to use inlet and outlet flow meters with higher accuracy and stability to monitor the increment of the mud pit.

(4) The feasibility and advantages of early monitoring of gas kick at the bottom of the riser have been demonstrated. A solution for monitoring gas kick at the riser using ultrasonic transmission monitoring method was proposed. Through experimental and numerical simulation research, the feasibility and effectiveness of ultrasonic transmission method were preliminarily explored. On this basis, a quantitative description method for wellbore gas kick degree based on gas kick monitoring data at the riser is proposed. This method is expected to accurately detect gas kick and conduct real-time quantitative evaluation of wellbore gas kick degree, thereby providing data support for the next reasonably well control. At present, multiple domestic research teams and petroleum companies have carried out experiments and field tests on ultrasonic monitoring of gas invasion in risers, confirming the effectiveness of this method. It is worth noting that on this basis, the author has proposed a new idea of moving the monitoring points downward to the near-bit area at the bottom of the well and integrating dielectric signals for collaborative monitoring of complex overflow. Relevant research progress can be referenced in the author's related journal papers.

Chapter 3
Risk Assessment Method of Natural Gas Hydrate Formation in the Wellbore of Deepwater Drilling

Abstract After gas influx occurs in deepwater drilling, a large amount of gas entering the annulus will form natural gas hydrates in an environment of high pressure and low temperature. There is a risk of blocking the choke and kill lines, subsea blowout preventers, wellheads, risers, etc., which poses a great threat to the normal progress of drilling operations and well control safety. In recent years, scholars at home and abroad have conducted certain research on this issue, mainly focusing on the prediction of formations containing natural gas hydrates in deepwater environments and the prediction of the formation areas of natural gas hydrates in the wellbore during deepwater drilling. At present, the prediction methods for formations containing natural gas hydrates mainly include BSR (bottom simulation reflection), the phase equilibrium condition method, logging in the geophysical field, and geochemical and geological methods, etc.; while the prediction of the formation areas of natural gas hydrates in the wellbore during deepwater drilling is mainly to predict the areas where hydrates may appear according to the temperature and pressure field in the wellbore and the phase equilibrium conditions of natural gas hydrates. It can be seen that the prediction of the formation of natural gas hydrates, especially the prediction of the formation of natural gas hydrates in the wellbore, mostly focuses on qualitative research, and there is a lack of quantification or evaluation of its formation risk, which is unfavorable for both drilling design and construction. The main reason is that it is difficult to quantitatively assess the relationship between various influencing factors and the hydrate formation risk. Therefore, by analyzing the influencing factors of hydrate formation in the wellbore of deepwater drilling, identifying the factors with great influence, uncertainty, or difficulty in quantitative assessment, seeking a method for quantitatively evaluating the hydrate formation risk in the wellbore, and then establishing a set of quantitative assessment methods for the natural gas hydrate formation risk in the wellbore of deepwater drilling is conducive to reducing or avoiding the drilling risks caused by the formation of hydrates in the wellbore during deepwater drilling, and has practical significance for deepwater drilling design and construction.

Y. Xu, *Assessment of Risks Related to Wellbore Safety in Deepwater Drilling*,
https://doi.org/10.1007/978-981-95-1668-1_3

3.1 Analysis of Influencing Factors for the Formation of Natural Gas Hydrates in the Wellbore

The formation process of hydrates in the wellbore mainly consists of four stages: gas dissolution, nucleation, growth, and stabilization. The nucleation of dissolved natural gas and water molecules needs to be formed under certain sub-cooling and pressure conditions through a certain induction time. Once the hydrate nuclei are formed, under suitable pressure and temperature conditions, natural gas hydrates will be formed after a certain growth time. It can be seen that the formation process of natural gas hydrates (such as the speed and quantity of formation) not only requires a high-pressure and low-temperature environment but also is related to sub-cooling, induction time, growth time, and so on. Among them, the influencing mechanisms of induction time, growth time, and others on the formation of hydrates are still not perfect, and relevant experiments need to be carried out in the laboratory to determine them. The obtained results are also numerical values under ideal conditions, which have a certain gap with the actual situation. Sub-cooling is one of the main driving forces in the formation process of natural gas hydrates. It comprehensively reflects the influence of factors such as the phase equilibrium conditions of hydrates, construction parameters, and environmental loads. It is a factor that has a great influence on the risk of hydrate formation and has the characteristics of parameters suitable for analyzing and evaluating the risk of hydrate formation in the wellbore during the drilling process. However, there is no clear quantitative relationship between sub-cooling and the risk of hydrate formation, and it is difficult to directly use it to quantitatively evaluate the risk of hydrate formation. Therefore, it is necessary to conduct research on this issue and seek a method to establish a quantitative characterization relationship between sub-cooling and the risk of hydrate formation to provide an effective means for the quantitative evaluation of the risk of hydrate formation in the wellbore.

According to the definition of sub-cooling, it is the difference between the wellbore annulus temperature at a certain depth in the wellbore and the phase equilibrium temperature of hydrates. Therefore, it is first necessary to calculate the phase equilibrium conditions of natural gas hydrates and the wellbore annulus temperature field.

3.1.1 Phase Equilibrium Conditions of Natural Gas Hydrates

At present, the commonly used methods for determining the phase equilibrium conditions of natural gas hydrates mainly include the graphical method, empirical formula method, equilibrium constant method, and molecular thermodynamic model method. Although the first three prediction methods are convenient and simple, the calculation errors are relatively large. The molecular thermodynamic model is a thermodynamic equilibrium equation for the formation of natural gas hydrates obtained based on thermodynamic theory:

$$\frac{\Delta\mu_0}{RT_0} - \int_{T_0}^{T_H} \frac{\Delta H_0 + \Delta C_K (T_H - T_0)}{RT_H^2} dT_H + \int_{P_0}^{P_H} \frac{\Delta V}{RT_H} dp_H = ln\frac{f_w}{f_{wr}} - \sum_{i=1}^{l} M_i ln\left(1 - \sum_{j=1}^{L} \theta_{ij}\right) \tag{3.1}$$

If an inhibitor is added:

$$\ln\frac{f_w}{f_{wr}} = \ln y_w x_w \tag{3.2}$$

where, $\Delta\mu_0$ is the chemical potential difference between the empty hydrate lattice and water in pure water under the standard state, J/mol; R is the universal gas constant, J/(mol·K); T_0 is the temperature under the standard state, K; T_H is the phase-state temperature of hydrate formation, K; ΔH_0 is the specific enthalpy difference between the empty hydrate lattice and pure water, J/kg; ΔC_K is the specific heat capacity difference between the empty hydrate lattice and pure water, J/(kg·K); p_H is the phase-state pressure of hydrate formation, Pa; p_0 is the pressure under the standard state, Pa; ΔV is the specific volume difference between the empty hydrate lattice and pure water, m^3/kg; f_w is the fugacity of water in the water-rich phase, Pa; f_{wr} is the fugacity of water at the reference state T_H and P_H, Pa; l is the number of types of hydrates; M_i is the number of type-i cavities in a unit water molecule of the hydrate phase; L is the number of components that can form hydrates; θ_{ij} is the occupancy rate of the type-j guest molecule in the type-i cavity; x_w is the molar fraction of water in the water-rich phase, dimensionless; y_w is the activity coefficient of water in the water-rich phase.

This model has a rigorous calculation and derivation process and high calculation accuracy, and can be used to calculate the phase equilibrium conditions of natural gas hydrates.

According to the analysis of relevant literature, the gas composition of natural gas hydrates and the salinity of pore water have a great influence on the phase equilibrium curve of natural gas hydrates. Therefore, when predicting whether a certain area of the formation or the wellbore contains natural gas hydrates, it is necessary to first sample and analyze the natural gas composition and water salinity in this area or wellbore to determine the corresponding phase equilibrium conditions of natural gas hydrates.

3.1.2 *Calculation of Subcooling Degree of Natural Gas Hydrate in Wellbore*

According to the definition of subcooling degree, by using the temperature–pressure field calculation model of deep-water drilling wellbore introduced in 2.1 of Chap. 2 and the phase-equilibrium conditions of natural gas hydrate in 3.1.1 of this chapter, the subcooling degree of natural gas hydrate in the wellbore at different well depths can be calculated. The area composed of all subcooling degrees is the area where natural gas hydrate in the wellbore is likely to form, and it has the conditions for the stable existence of hydrate. That is, if hydrate is formed, it must be in this area.

The following takes a deep-water well as an example for calculation. The water depth of this well is 1524 m, the open-hole section is 660 m, the outer diameter of the riser is 508 mm, the inner diameter is 493 mm, the displacement is 28 L/s, the inlet temperature of the drilling fluid is 20 °C, the geothermal gradient is 2.7 °C/100 m, the sea-water surface temperature is 20 °C, the density of the drilling fluid is 1.07 g/cm^3 (seawater), the salinity of seawater is 3.5%, and the main components of shallow-layer gas are methane (90%), ethane (7%) and propane (3%). Then the wellbore temperature field and the phase-equilibrium curve of natural gas hydrate (the temperature–pressure curve of natural gas hydrate phase equilibrium needs to be converted into a temperature-well depth curve by combining with the wellbore annulus pressure field) are shown in Fig. 3.1.

In Fig. 3.1, the area enclosed by the wellbore annulus temperature curve, the hydrate phase-equilibrium temperature-depth curve, and the bottom of the well is

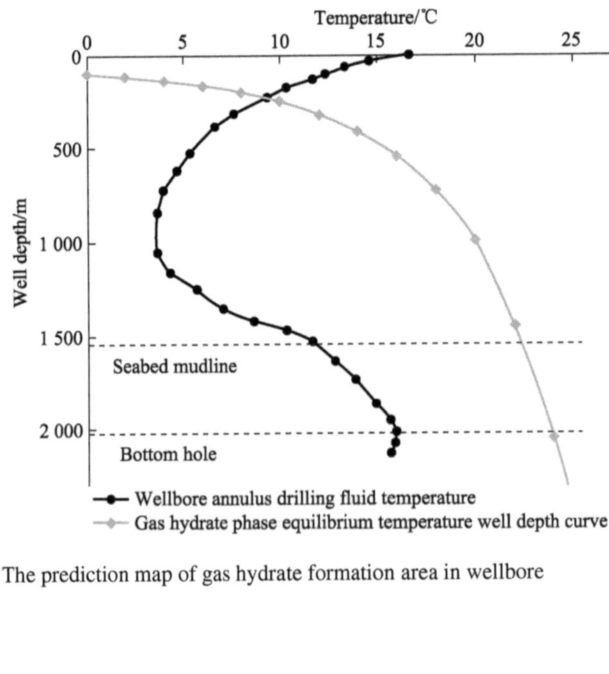

Fig. 3.1 The prediction map of gas hydrate formation area in wellbore

the area where natural gas hydrates may exist. It can be seen that the sub-cooling degree at each well depth changes significantly.

It can also be seen from Fig. 3.1 that although the inlet temperature of the drilling fluid and the surface temperature of the seawater are both 20 °C, due to the great water depth and the relatively shallow depth of the drilled formation, the temperature of the drilling fluid in the wellbore annulus is affected far more by the low temperature at the seabed than by the heat transfer of the geothermal temperature. As a result, the drilling fluid in the entire annulus is in a low-temperature state, which greatly increases the formation area of natural gas hydrates (sub-cooling degree). However, when the well depth is deeper, as the well depth increases, the heat transfer time of the geothermal temperature increases significantly, so the temperature of the drilling fluid in the annulus will increase significantly, and the formation area of natural gas hydrates will correspondingly decrease significantly. At the same time, due to the change in the wellbore temperature–pressure field at this time, the natural gas hydrate phase-equilibrium temperature-depth curve will also change. Taking the well with a depth of 4524 m (including a water depth of 1524 m, a cemented section of 1200 m, and an open-hole section of 1800 m) as an example, the calculation results are shown in Fig. 3.2.

As can be seen from Fig. 3.2, when the well depth is relatively deep, the temperature-well depth curve of natural gas hydrate phase equilibrium shifts downward, and the temperature curve of the wellbore annulus shifts to the right, especially below the mud line. When the well depth is 2184 m, the area from 300 to 2184 m

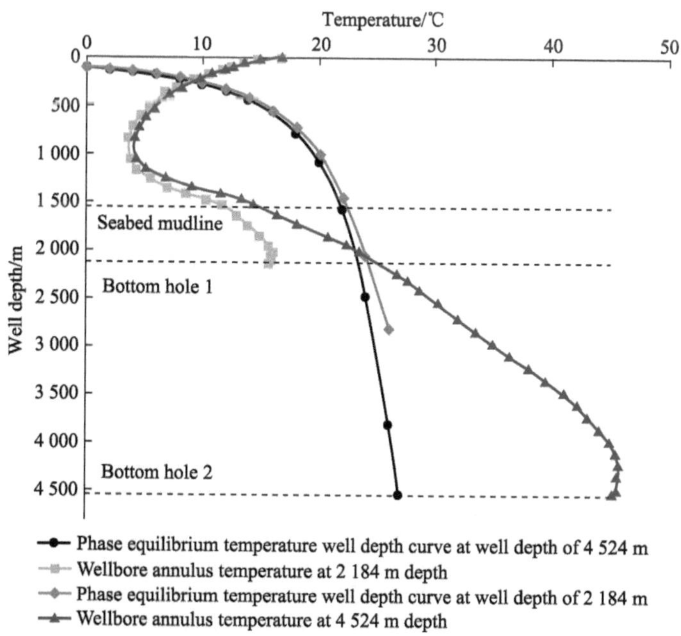

Fig. 3.2 Comparison of gas hydrate formation area with different depth

(bottom of the well) is the generation area of natural gas hydrates, and the degree of subcooling (the horizontal span of the generation area) is relatively large. When the well depth is 4524 m, the area from 350 to 1850 m is the generation area of natural gas hydrates, and the degree of subcooling is smaller than that in shallow-layer drilling. This indicates that in deep-water drilling, natural gas hydrates are more likely to form in shallow-layer formation drilling than in deep-layer formation drilling. That is, when drilling in shallow-layer formations, the length of the well section where natural gas hydrates form in the wellbore and the associated risks increase significantly compared to deep-layer drilling.

3.1.3 Quantitative Characterization Relationship Between Subcooling Degree and Hydrate Formation Risk

In order to use the subcooling degree to quantitatively describe the hydrate formation risk at different well depths, the concept of "subcooling degree density" can be introduced.

The "subcooling degree density" is defined as the area enclosed by the hydrate formation phase equilibrium curve and the annulus temperature curve within a unit depth range, and can be calculated according to the following formula:

$$\rho_{\mathrm{scd}} = \int_{h}^{h+\Delta h} (T_{\mathrm{ba}} - T_{\mathrm{te}})\mathrm{d}h \tag{3.3}$$

where, ρ_{scd} is the subcooling degree density, °C · m; T_{ba} is the phase equilibrium temperature of natural gas hydrate at a certain depth, °C; T_{te} is the temperature in the wellbore at a certain depth, °C; Δh is the unit depth, m, and its value is determined according to the specific situation.

The subcooling degree density reflects the possibility of the formation of natural gas hydrates within a certain depth range. It comprehensively reflects the influence of construction parameters, environmental loads, and the phase equilibrium conditions of hydrates, and has the function of quantitative risk assessment.

Figure 3.3 is the subcooling degree density-depth curve calculated and drawn according to the annulus temperature curve and the phase equilibrium temperature-depth curve when the well depth is 4524 m in Fig. 3.2, where Δh is taken as 200 m. The safety lower limit ρ_{scd}^* of the subcooling degree density can be set according to the actual situation. It is considered that if the subcooling degree density at a certain well depth exceeds ρ_{scd}^*, there is a risk of hydrate formation, and the magnitude of the risk is:

$$R_{\mathrm{scd}} = \frac{\rho_{\mathrm{scd}} - \rho_{\mathrm{scd}}^*}{\rho_{\mathrm{scd}}^*} \tag{3.4}$$

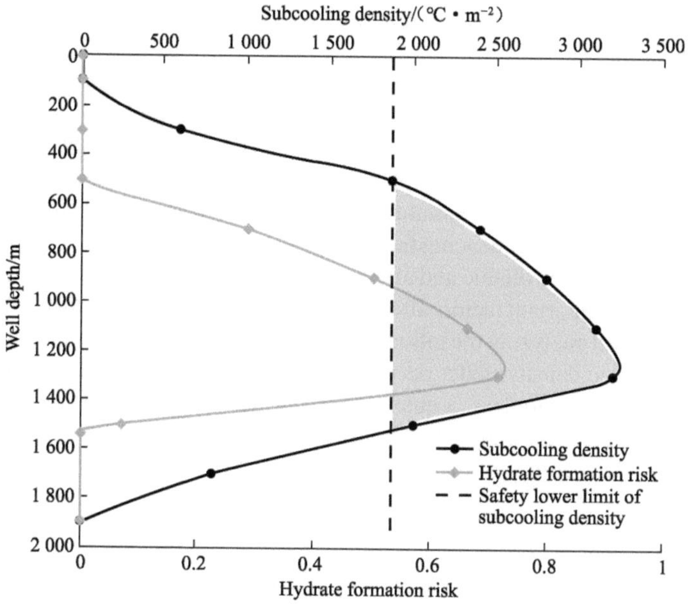

Fig. 3.3 Subcooling density-depth curve and gas hydrate formation risk distribution

Thus, a quantitative characterization relationship between the subcooling degree and the hydrate formation risk is established.

As shown in Fig. 3.3, based on the subcooling degree, a quantitative risk analysis of the natural gas hydrate formation risk in the entire wellbore can be carried out through this figure. It can be seen that the red area is the hydrate formation risk area. The hydrate formation risk values near the subsea wellhead and in the range of 1000 m to 1400 m of the riser are the largest, which are high-risk areas for natural gas hydrate formation. This is mainly due to the influence of the high-pressure and low-temperature environment in these areas. At the same time, this figure also reflects the change of the hydrate formation risk with the well depth: the hydrate formation risk in the seawater section gradually increases with the water depth, starts to decline after reaching the peak, and the hydrate formation risk basically disappears after entering the formation section.

3.2 Quantitative Evaluation Method of Hydrate Formation Risk Based on Construction Parameters

Through the previous analysis, the subcooling degree is one of the main driving forces for hydrate formation. If hydrates are formed in the wellbore, it must be within the area where the subcooling degree is greater than zero (the predicted area for hydrate formation). Therefore, it is reliable to analyze the formation risk based on the subcooling degree.

In 3.1.3, a quantitative characterization relationship between the subcooling degree and the hydrate formation risk in the wellbore was established by introducing the subcooling degree density. Therefore, all factors that affect the subcooling degree density (that is, factors affecting the phase equilibrium curve and the temperature and pressure field in the wellbore annulus) will affect the formation risk of natural gas hydrates. Among many influencing factors, construction parameters are the main data that are artificially controllable and affect whether the drilling can proceed normally, and they are also important factors affecting the subcooling degree density. Therefore, by calculating and analyzing the influence of different construction parameters on the subcooling degree density under deepwater conditions, and making each construction parameter dimensionless, a quantitative risk assessment can be carried out for the area with a relatively high risk of hydrate formation in the wellbore to ensure the applicability and accuracy of the assessment results.

3.2.1 Analysis of the Influence of Construction Parameters on the Formation Area of Natural Gas Hydrates Under Deepwater Conditions

Taking the deepwater well mentioned in 3.1.2 as an example, calculate the formation area of natural gas hydrates (the area enclosed by the wellbore annulus temperature curve and the hydrate phase equilibrium temperature curve) under different construction parameter conditions of this well, that is, the area where the subcooling degree is greater than zero.

(1) Influence of Circulation Displacement

Calculate the formation areas of natural gas hydrates when the circulation displacements are 28 L/s, 36 L/s, and 45 L/s respectively, as shown in Fig. 3.4.

(2) Influence of Drilling Fluid Density

Calculate the formation areas of natural gas hydrates when the densities of the drilling fluid are 1.07 g/cm^3 and 1.5 g/cm^3 respectively, as shown in Fig. 3.5.

(3) Influence of Thermal Conductivity of Drilling Fluid

Calculate the formation areas of natural gas hydrates when the thermal conductivities of the drilling fluid are 0.45 W/(m·°C), 1.0 W/(m·°C), and 1.8 W/(m·°C) respectively, as shown in Fig. 3.6.

(4) Influence of Specific Heat of Drilling Fluid

Calculate the formation areas of natural gas hydrates when the specific heats of the drilling fluid are 1.67 J/(g·°C), 2.5 J/(g·°C), and 4.0 J/(g·°C) respectively, as shown in Fig. 3.7.

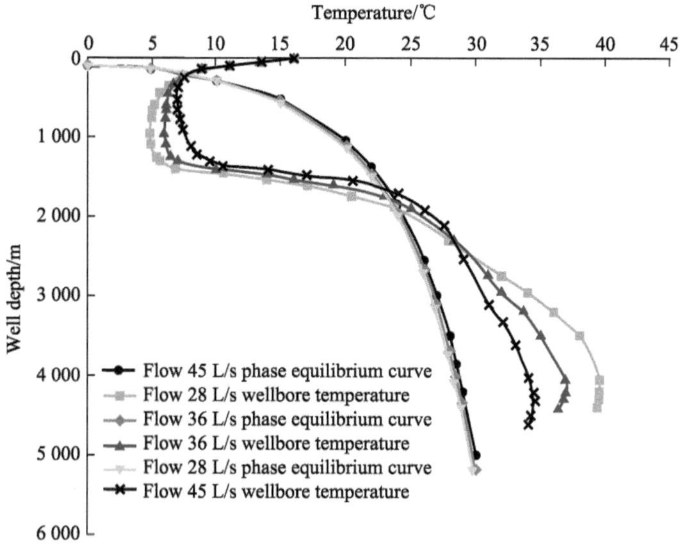

Fig. 3.4 The effects of different displacement on gas hydrate formation area

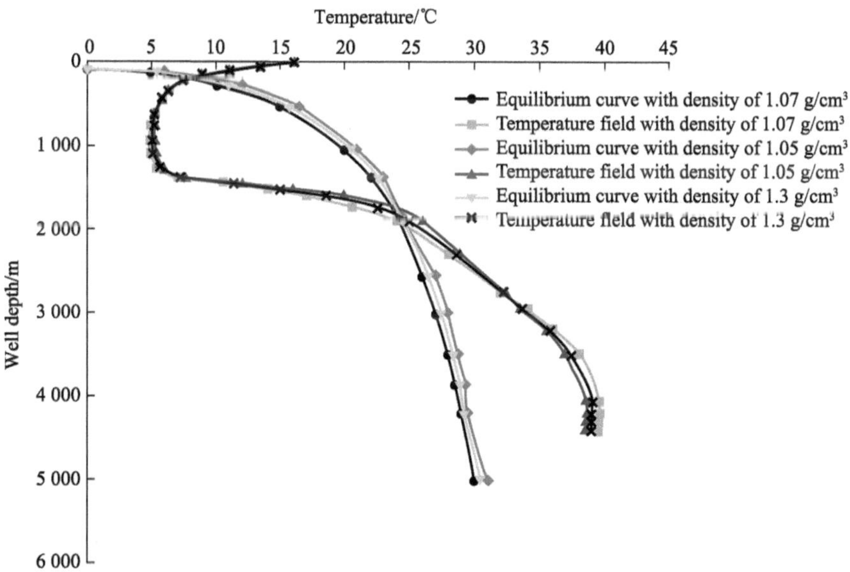

Fig. 3.5 The effects of different drilling fluid density on gas hydrate formation area

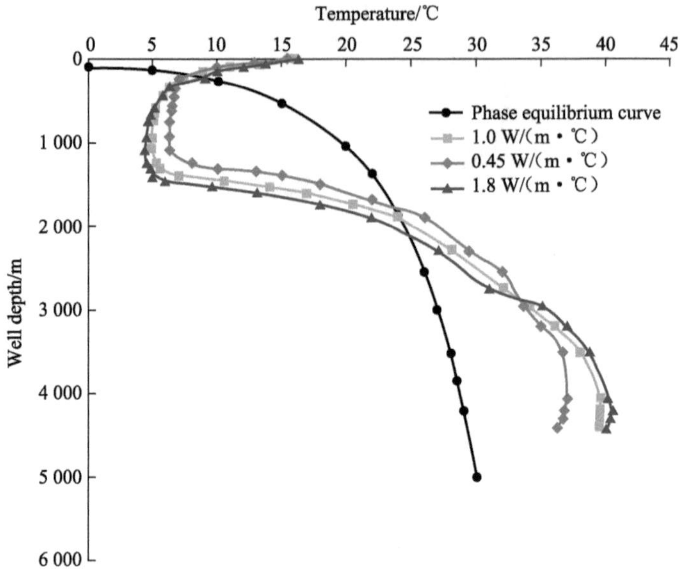

Fig. 3.6 The effects of different drilling fluid thermal conductivity on gas hydrate formation area

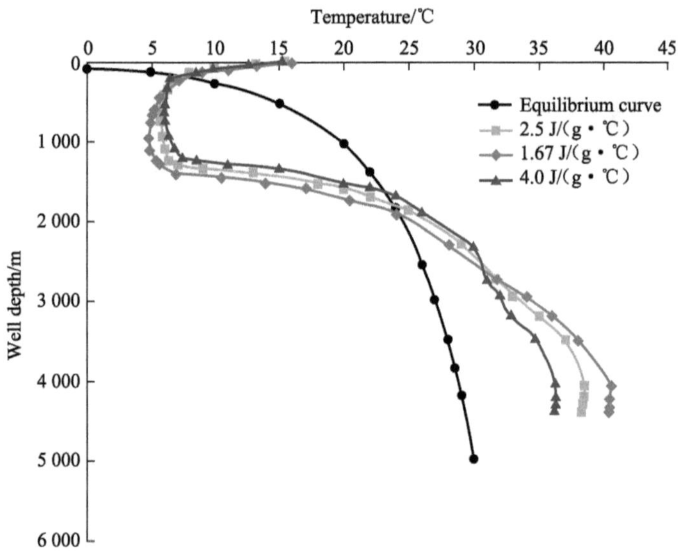

Fig. 3.7 The effects of different drilling fluid specific heat on gas hydrate formation area

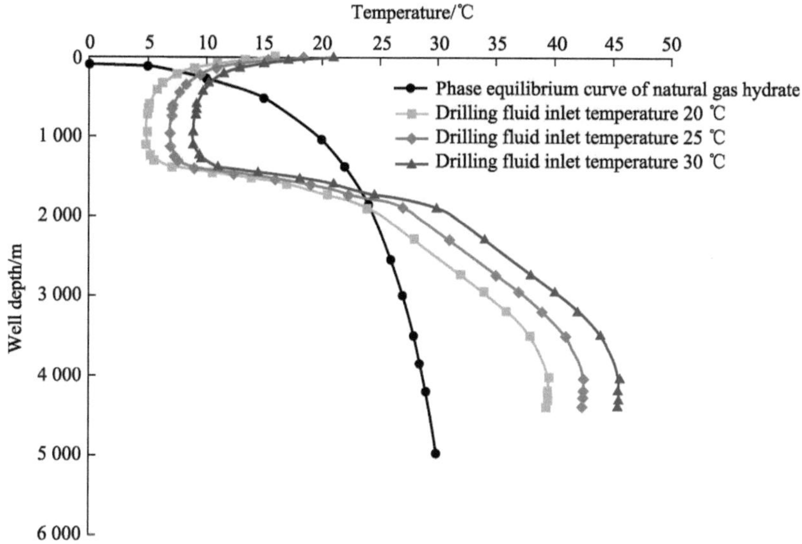

Fig. 3.8 The effects of different drilling fluid inlet temperature on gas hydrate formation area

(5) Influence of Inlet Temperature of Drilling Fluid

Calculate the formation areas of natural gas hydrates when the inlet temperatures of the drilling fluid are 20 °C, 25 °C, and 30 °C respectively, as shown in Fig. 3.8.

(6) Influence of NaCl Concentration

Calculate the formation areas of natural gas hydrates when the NaCl concentrations at the inlet of the drilling fluid are 0, 6%, and 9% respectively, as shown in Fig. 3.9.

(7) Influence of Pump Shutdown Time

Calculate the formation areas of natural gas hydrates when the pump shutdown times are 0, 15 min, and 30 min respectively, as shown in Fig. 3.10.

3.2.2 *Dimensionless Analysis of the Influence of Construction Parameters on the Subcooling Degree Density*

According to the analysis in 3.1.3 and 3.2.1, the subcooling degree density is the largest near the subsea wellhead and the lower part of the riser, and the risk of natural gas hydrate formation is relatively high. Moreover, this part is most affected by the construction parameters, so it is necessary to conduct a key analysis of the risk in this area.

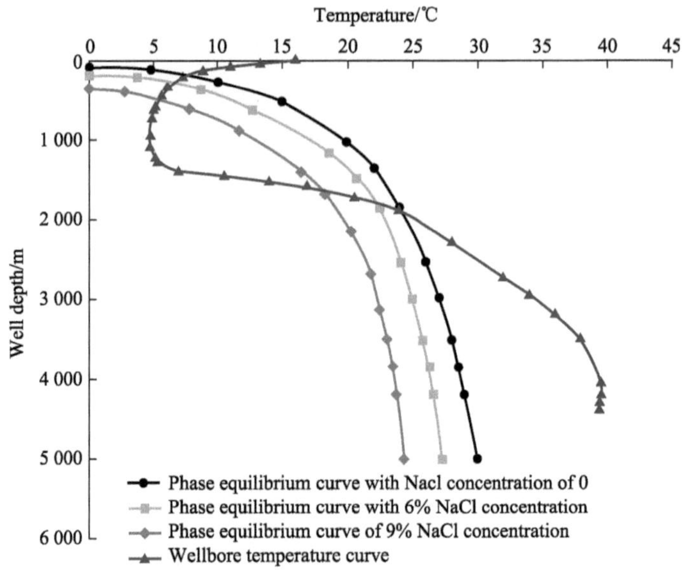

Fig. 3.9 The effects of different concentration of NaCl on gas hydrate formation area

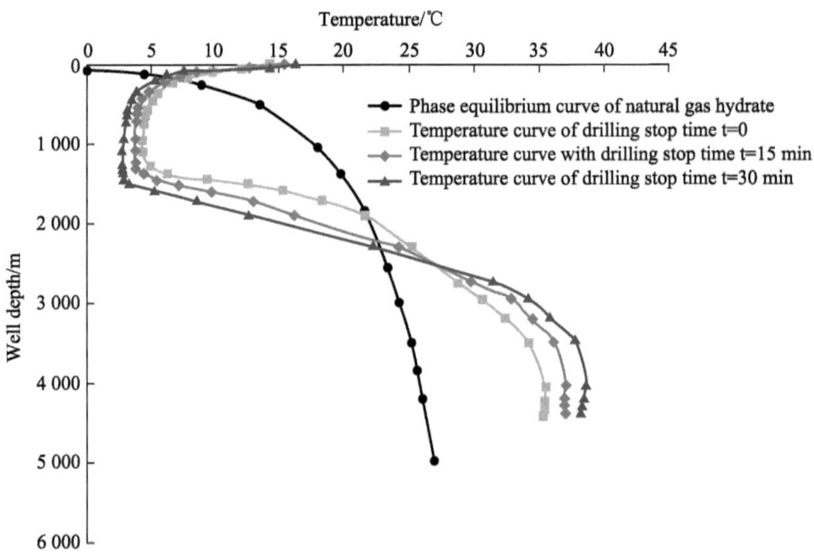

Fig. 3.10 The effects of different stop drilling time on gas hydrate formation area

Taking the deepwater well in 3.1.2 as an example, based on the definition of the subcooling degree density in 1.3, the part with $h = 700$ m above the subsea wellhead (from 800 to 1500 m) is selected to calculate the subcooling degree density under different construction parameter conditions. The calculation results are shown in Table 3.1.

Since the units and value intervals of various construction parameters are different, they are not comparable to each other, and it is impossible to quantitatively describe the magnitude of the influence of each construction parameter on the subcooling degree density. Therefore, a dimensionless treatment is required.

Taking the influence of the inlet temperature of the drilling fluid on the subcooling degree density as an example, first divide the inlet temperature of the drilling fluid and the subcooling degree density under the conditions of 25 °C and 30 °C by the corresponding parameters under the initial condition (20 °C). The calculation results are shown in Table 3.2.

Secondly, calculate the sensitive factor γ_T, that is, the α_T required to change the unit $\beta_{\rho scd}$:

$$\gamma_T = \frac{\alpha_{T2} - \alpha_{T1}}{\beta_{pscd2} - \beta_{pscd1}} \tag{3.5}$$

Substitute the data to get $\gamma_T = -2.055$

The negative sign represents that as the inlet temperature of the drilling fluid increases, the subcooling degree density decreases.

Calculate the γ_i of other construction parameters according to this method, and the results are shown in Table 3.3.

The sensitive factor γ_i in Table 3.3 is calculated based on the initial state (the NaCl concentration is 6% and the pump shutdown time is 15 min). Its numerical value represents the dimensionless quantity of the construction parameter required to change the unit subcooling degree density within the range of $\Delta\ h = 700$ m. The smaller the value, the greater the influence of the construction parameter on the subcooling degree density, and vice versa. A positive γ_i indicates that the subcooling degree density increases as the construction parameter increases, and a negative γ_i indicates that the subcooling degree density decreases as the construction parameter increases.

Through the above analysis, among the many construction parameters of this deepwater well, the order of the influence on the subcooling degree density from large to small is: the thermal conductivity of the drilling fluid, the inlet temperature of the drilling fluid, the displacement, the NaCl concentration, the density of the drilling fluid, the pump shutdown time, and the specific heat of the drilling fluid. Among them, the thermal conductivity of the drilling fluid, the inlet temperature, the displacement, and the NaCl concentration have the greatest influence on the subcooling degree density. Therefore, in order to reduce the risk of natural gas hydrate formation in the wellbore, it is necessary to focus on controlling these four parameters. According to the value of γ_i and comprehensively considering the actual situation, the thermal

Table 3.1 The subcooling density under different conditions of construction parameters

Construction parameters							Subcooling density ($10^2 \cdot °C \cdot m$)
Displacement/($L \cdot s^{-1}$)	Drilling fluid density (g/cm^3)	Drilling fluid thermal conductivity ($W/m \cdot °C$)	Drilling fluid specific heat /($J \cdot g^{-1} \cdot °C^{-1}$)	Inlet temperature of drilling fluid /°C	NaCl concentration /%	Pump-stop time /min	
28	1.07	1.80	1.67	20	0	0	89
36	1.07	1.80	1.67	20	0	0	82
45	1.07	1.80	1.67	20	0	0	70
28	1.30	1.80	1.67	20	0	0	92
28	1.50	1.80	1.67	20	0	0	96
28	1.07	1.00	1.67	20	0	0	87
28	1.07	0.45	1.67	20	0	0	73
28	1.07	1.80	2.50	20	0	0	83
28	1.07	1.80	4.00	20	0	0	72
28	1.07	1.80	1.67	25	0	0	78
28	1.07	1.80	1.67	30	0	0	65
28	1.07	1.80	1.67	20	6	0	75
28	1.07	1.80	1.67	20	9	0	50
28	1.07	1.80	1.67	20	9	15	96
28	1.07	1.80	1.67	20	9	30	107

Table 3.2 The dimensionless of drilling fluid inlet temperature

Inlet temperature of drilling fluid /°C	Subcooling density $(10^2 \cdot °C \cdot m)$	Dimensionless inlet temperature of drilling fluid α_T	Dimensionless subcooling density $\beta_{\rho scd}$
20	89	None	None
25	78	1.25	0.876
30	65	1.50	0.730

Table 3.3 The sensitive factor γ_i corresponding to different construction parameters

Construction parameters	Drilling fluid density	Drilling fluid thermal conductivity	Drilling fluid specific heat	Inlet temperature of drilling fluid	NaCl concentration	Pump-stop time	Drilling fluid density
γ_i	−2.38	4.16	1.95	−7.27	−2.06	−2.89	6.49

conductivity of the drilling fluid should be reduced, and the inlet temperature of the drilling fluid, the displacement, and the NaCl concentration should be increased.

3.3 Risk Assessment Method of Natural Gas Hydrate Formation Based on Analytic Hierarchy Process

In 3.2, a quantitative evaluation method of hydrate formation risk based on construction parameters was established. According to the evaluation results, the construction parameters during the drilling process can be reasonably regulated to reduce or even avoid the hydrate formation risk. However, in addition to the quantitatively controllable construction parameters, there are many factors affecting the formation of natural gas hydrates in the wellbore. Among them, there are some factors that are difficult to determine or quantitatively describe, and many factors are internally related, making it difficult to accurately calculate or evaluate them through a mathematical model. Expert experience can be fully utilized to comprehensively assess the formation risk. On this basis, considering the factors affecting the formation of natural gas hydrates, the analytic hierarchy process is used to conduct a semi-quantitative assessment of the risk.

3.3.1 Qualitative Risk Assessment Based on the Expert Scoring Method

The expert scoring method is introduced to conduct a qualitative risk assessment of the natural gas hydrate formation risk in the wellbore:

First, predict the formation area of natural gas hydrates in the wellbore (where the subcooling degree is greater than zero) by the method in 3.1.2, and determine the subcooling degree at different well depths.

Then, according to the specific situation of the formation at this well location, the drilling design and construction, and the relevant equipment, divide the predicted area into several groups along the well depth direction, assuming that each group has the same risk.

Experts, combining with the experience of the area or adjacent wells, score the natural gas hydrate formation risk of each group according to different subcooling degrees, and process the scoring data to determine the risk level.

Finally, draw the relationship curve between the natural gas hydrate formation risk and the well depth according to the risk level.

The corresponding relationship between the risk level and the scoring value is shown in Table 3.4. Taking a certain deepwater drilling in 3.2 as an example, the scoring situation of 5 experts is shown in Table 3.5.

Draw the relationship curve between the formation risk of natural gas hydrates and the well depth according to the corresponding relationship between the risk levels in Table 3.4 and the well depth, as shown in Fig. 3.11.

Through Fig. 3.11, a qualitative understanding of the natural gas hydrate formation risk in the entire wellbore can be obtained. It can be seen that the risk level at the subsea wellhead and near the 1000 m of the riser is 4, which is a high-risk area for natural gas hydrate formation. This is mainly the result of a series of factors, such as the high-pressure and low-temperature environment at the subsea wellhead, the special structure of the equipment, the high-pressure and low-temperature environment of the riser, and the time required for the formation of natural gas hydrates. Based on the prediction of the natural gas hydrate formation area, this result comprehensively reflects different subcooling degrees, different formation or drilling working conditions, data of adjacent wells, and the on-site experience of various experts, and

Table 3.4 The relationship between risk level and score value

Score (The higher the score, the greater the risk)	Risk level	Explanation
0	0	No risk
1–3	1	Essentially no risk
3–5	2	Slight risk
5–7	3	Moderate risk
7–9	4	High risk

Table 3.5 Expert scoring and risk rating

Well depth /m	Expert 1	Expert 2	Expert 3	Expert 4	Expert 5	Average score	Risk level
246(Regional vertex)	0	0	0	0	0	0	0
300–500	2	3	1	2	2	2	1
500–800	6	5	6	7	5	5.8	3
800–1300	7	8	8	8	9	8	4
1300–1500	7	8	6	6	7	6.8	3
1524 (Subsea wellhead)	7	9	9	8	8	8.2	4
1600–1800	6	5	6	6	7	6	3
1800–2184	3	4	4	3	3	3.4	2

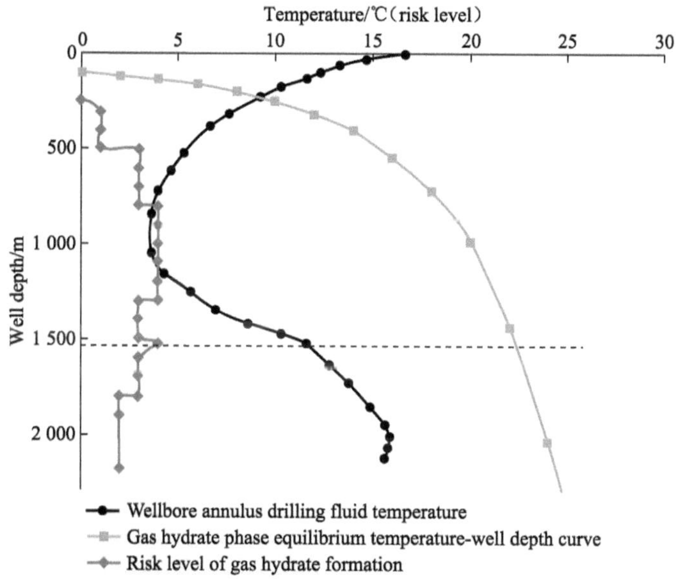

Fig. 3.11 Risk level distribution of natural gas hydrate

has practical guiding significance for indicating the high-risk areas of natural gas hydrates in the wellbore.

3.3.2 Semi-quantitative Risk Assessment Based on Analytic Hierarchy Process

On the basis of dividing the risk levels of the predicted areas for hydrate formation in 3.3.1, the analytic hierarchy process can be used to comprehensively analyze many factors affecting the hydrate formation risk in the wellbore, and then evaluate the relative magnitudes of the influence of each factor on the hydrate formation risk.

The hierarchical structure model of the natural gas hydrate formation risk in deepwater drilling is shown in Fig. 3.12.

(1) Construction of the Judgment Matrix

Based on the influencing factors presented in the risk hierarchical structure model in Fig. 3.12, the 9-level scaling method is used to conduct pairwise comparisons of the relative importance between any one factor and other factors. Thus, the judgment matrix A can be constructed:

$$A = \begin{bmatrix} u_{11} & \cdots & u_{1n} \\ \vdots & \vdots & \vdots \\ u_{n1} & \cdots & u_{nn} \end{bmatrix}$$

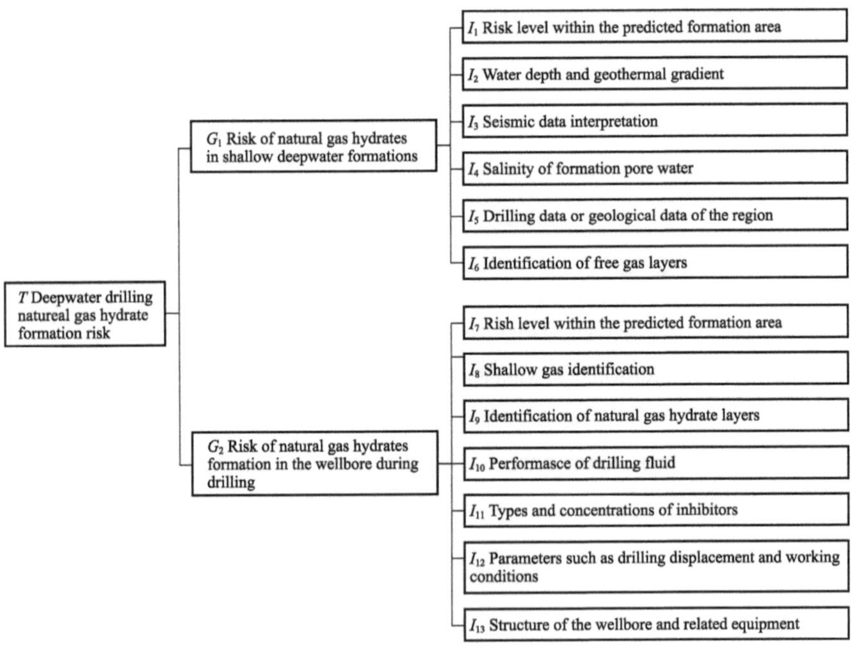

Fig. 3.12 Risk hierarchy model of natural gas hydrate in deep water drilling

In the matrix, u_{ij} represents the importance degree of factor I_i relative to I_j. The numerical values are determined by the expert group through pairwise comparisons of the factors influencing the natural gas hydrate formation risk, which reflects the influence proportion of each factor on the natural gas hydrate formation in deep-water drilling. In this paper, three judgment matrices can be constructed, namely: $T - G_i$, $G_1 - I_i$, and $G_2 - I_i$. The meanings of the 9-level scaling are shown in Table 3.6.

(2) Calculation of the Relative Weights of Factors

The eigenvalue method is used to find the maximum eigenvalue $\lambda_{\max}$ of the judgment matrix A, that is, $AW = \lambda_{\max} W$, then $W = \left(w_1, w_2, \cdots, w_n\right)$ is the relative weight vector of A, and w_i is the weight for factor ranking.

(3) Consistency Check

Since the judgment matrix is obtained by the expert group through a certain degree of fuzzy quantification based on their own experience, there will be certain deviations in the evaluations of individual experts. Therefore, the construction of the judgment matrix cannot be completely consistent, and it is necessary to introduce the consistency ratio *CR* to determine whether it satisfies the consistency.

$$CR = \frac{CI}{RI} \tag{3.6}$$

where *RI* is the proportionality coefficient, which is related to the order n of the judgment matrix, and the specific values are shown in Table 3.7.

The expression of the consistency index *CI* is:

$$CI = \frac{\lambda_{\max} - n}{n - 1} \tag{3.7}$$

Table 3.6 The meanings of 9 levels

Importance of u_i compared to u_j	Scale
The two elements are equally important	1
The former is slightly more important than the latter	3
The former is obviously more important than the latter	5
The former is strongly more important than the latter	7
The former is extremely more important than the latter	9
Reverse comparison	1/3, 1/5, 1/7, 1/9
Intermediate values of the above—adjacent judgments	2, 4, 6, 8, 1/2, 1/4, 1/6, 1/8

Table 3.7 The evaluation value of consistency stochastic (RI)

1	2	3	4	5	6	7	8	9
0	0	0.58	0.90	1.12	1.24	1.32	1.41	1.45

Table 3.8 The weight of single layer $T - G_i$ sorting index calculation results and the test of consistency

T	G_1	G_2	W
G_1	1	1/2	0.3333
G_2	2	1	0.6667

$\lambda_{max} = 2\lambda_{max} = 2$, $CI = 0$, $CR = 0 < 0.1$, meet the consistency condition

Through calculation, if $CR < 0.1$, it is considered that the consistency requirement is met; otherwise, the judgment matrix needs to be appropriately revised.

Taking a deep—water drilling case in 3.2 as an example for the risk assessment of natural gas hydrate formation in deep—water drilling, the calculation results of each index value and the consistency test values of the judgment matrices $T - G_i$, $G_1 - I_i$, and $G_2 - I_i$ are presented. See Tables 3.8, 3.9, and 3.10.

According to the calculation results of the index weights in the above single-layer ranking, calculate the overall ranking of the influence degrees of all factors on the natural gas hydrate formation risk in deepwater drilling, as shown in Table 3.11.

Table 3.9 The weight of single layer $G_1 - I_i$ sorting index calculation results and the test of consistency

G_1	I_1	I_2	I_3	I_4	I_5	I_6	W
I_1	1	3	5	3	2	5	0.3841
I_2	1/3	1	2	1	3	4	0.1876
I_3	1/5	1/2	1	1/2	1	3	0.1002
I_4	1/3	1	2	1	3	3	0.1769
I_5	1/2	1/3	1	1/3	1	1/2	0.0818
I_6	1/5	1/4	1/3	1/3	2	1	0.0693

$\lambda_{max} = 2$, $CI = 0.09586$, $RI = 1.24$, $CR = 0.0773 < 0.1$, meet the consistency condition

Table 3.10 The weight of single layer $G_2 - I_i$ sorting index calculation results and the test of consistency

G_2	I_7	I_8	I_9	I_{10}	I_{11}	I_{12}	G_{13}	W
I_7	1	1/3	1/2	3	2	2	3	0.1466
I_8	3	1	3	5	4	3	5	0.3591
I_9	2	1/3	1	3	3	3	4	0.2081
I_{10}	1/3	1/5	1/3	1	1	2	3	0.0866
I_{11}	1/2	1/4	1/3	1	1	2	2	0.0863
I_{12}	1/2	1/3	1/3	1/2	1/2	1	2	0.0691
I_{13}	1/3	1/5	1/4	1/3	1/2	1/2	1	0.0442

$\lambda_{max} = 7.3221$, $CI = 0.05368$, $RI = 1.32$, $CR = 0.04068 < 0.1$, meet the consistency condition

Table 3.11 The total sequence of evaluation indexes

Hierarchy	G_1	G_2	Synthetic weight	Overall ranking
	0.3333	0.6667		
I_1	0.3841	0	0.128	3
I_2	0.1876	0	0.0625	5
I_3	0.1002	0	0.0334	10
I_4	0.1769	0	0.059	6
I_5	0.0818	0	0.0273	12
I_6	0.0693	0	0.0231	13
I_7	0	0.1466	0.0977	4
I_8	0	0.3591	0.2394	1
I_9	0	0.2081	0.1387	2
I_{10}	0	0.0866	0.0577	7
I_{11}	0	0.0863	0.0575	8
I_{12}	0	0.0691	0.0461	9
I_{13}	0	0.0442	0.0295	11

According to the overall ranking results in Table 3.11, in the example of this paper, the ranking of the influence of the risk of natural gas hydrates in the shallow deepwater formation G_1 and the risk of natural gas hydrate formation in the wellbore during deepwater drilling G_2 on the natural gas hydrate formation risk T in deepwater drilling is as follows: the risk of natural gas hydrate formation in the wellbore during shallow deepwater drilling (0.6667) > the risk of natural gas hydrates in the shallow deepwater formation (0.3333). The ranking of the factors affecting the natural gas hydrate formation risk T in deepwater drilling is: Shallow Gas Identification I_8 > Natural Gas Hydrate Layer Identification I_9 > Risk Level I_1 (Formation) within the Predicted Formation Area > Risk Level I_7 (Wellbore) within the Predicted Formation Area > Water Depth and Geothermal Gradient I_2 > Salinity of Formation Pore Water I_4 > Drilling Fluid Performance I_{10} > Type and Concentration of Inhibitor I_{11} > Parameters and Working Conditions such as Drilling Displacement I_{12} > Seismic Data Interpretation I_3 > Structure of Wellbore and Related Equipment I_{13} > Regional Drilling Data or Geological Data I_5 > Free Gas Layer Identification I_6.

According to the above ranking results, it can be seen that the five factors that have the greatest influence on the natural gas hydrate formation risk T in deepwater drilling in this example are: Shallow Gas Identification I_8 > Natural Gas Hydrate Layer Identification I_9 > Risk Level I_1 (Formation) within the Predicted Formation Area > Risk Level I_7 (Wellbore) within the Predicted Formation Area > Water Depth and Geothermal Gradient I_2. Among them, the Risk Level I_1 (Formation) within the Predicted Formation Area and the Risk Level I_7 (Wellbore) within the Predicted Formation Area can be calculated by the method in 3.1. The water depth and geothermal gradient are uncontrollable factors, and the Shallow Gas Identification I_8 and Natural Gas Hydrate Layer Identification I_9 depend on the advancement of the equipment and the abundance of regional data, which need to be judged according to the actual situation.

In field practice, the drilling design can be optimized and construction remedial measures can be taken according to the ranking of each influencing factor and its weight value obtained by this method, so as to reduce or avoid the risk of natural gas hydrate formation in deepwater drilling.

3.4 Summary

(1) By analyzing the influencing factors of the formation of natural gas hydrates in the wellbore, it is found that the subcooling degree has the parameter characteristics for analyzing and evaluating the hydrate formation risk in the wellbore during the drilling process. Through introducing the concept of subcooling degree density, a quantitative characterization relationship between the subcooling degree and the hydrate formation risk is established.

(2) Through the risk analysis based on the subcooling degree density, it is found that the risk of hydrate formation in the wellbore during shallow deepwater drilling is greater than that during deep drilling, and the high-risk areas are concentrated in the well sections of the subsea wellhead and the lower part of the riser.

(3) By calculating the subcooling degree density under different construction parameter conditions and using the dimensionless analysis, a quantitative evaluation method of hydrate formation risk based on construction parameters is established, which can quantitatively evaluate the influence of each construction parameter on the hydrate formation risk. For the deepwater well cited in this paper, in order to reduce the risk of natural gas hydrate formation in the wellbore, the thermal conductivity of the drilling fluid should be reduced according to the sensitive factor and the actual situation, and the inlet temperature of the drilling fluid, the displacement, and the NaCl concentration should be increased.

(4) In order to comprehensively consider the influence of other factors, based on the analytic hierarchy process and combined with the risk level division within the predicted hydrate formation area, a semi-quantitative evaluation method of hydrate formation risk in the wellbore is established. This method comprehensively considers the influence of expert on-site experience and other factors. In field practice, according to the ranking of each influencing factor and its weight obtained by this method, the drilling design can be optimized and construction remedial measures can be taken to reduce or avoid the risk of natural gas hydrate formation in the wellbore during deepwater drilling.

Chapter 4
Risk Assessment Method of Wellbore Structure in Deepwater Drilling Process

Abstract The problem of narrow safety density windows in deepwater drilling has always been one of the challenges faced in wellbore structure design and safe construction. Moreover, with the increase in water depth, the problem of narrow safety density windows becomes more prominent. To this end, at home and abroad, new equipment such as advanced deepwater drilling platforms and new process technologies such as underbalanced drilling, managed pressure drilling, aerated drilling, and dual-gradient drilling, as well as risk assessment technologies such as pre-drilling wellbore structure design and risk assessment methods based on probability and statistics have been adopted, which have solved this problem to a certain extent. Among them, risk assessment technology can quantitatively evaluate engineering risks based on theories such as risk analysis on the basis of existing equipment and processes, which is very necessary for high-risk and high-investment deepwater drilling and is conducive to achieving the goal of safe and efficient deepwater drilling. Through research, the pre-drilling wellbore structure design and risk assessment method based on probability and statistics uses logging data or seismic data of adjacent wells to probabilistically predict the safety density window of the target well, and quantitatively evaluates engineering risks such as leakage, blowout, collapse, and sticking of the existing wellbore structure design scheme before drilling, and then improves it. This method is effective for wells with abundant adjacent well data. However, with the continuous increase in the water depth of deepwater drilling and the drilling of unproven blocks, the abundance of adjacent well data is greatly limited, thus reducing the accuracy of wellbore structure risk assessment technologies such as regional probability and statistics methods. At the same time, there are currently no good methods or technologies for the quantitative evaluation of wellbore structure and engineering risks during deepwater drilling. The methods of accident statistics are mostly used to establish accident tree models, bow-tie models, safety barrier models, etc. and combine drilling conditions to macroscopically evaluate the wellbore structure and engineering risks during the drilling process, lacking a mechanistic understanding of engineering risks. With the gradual maturity and development of measurement while drilling and transmission technologies, the formation pressure and rock mechanical parameters of the drilled section, which are uncertain parameters before drilling, can be accurately determined during the drilling process.

Y. Xu, *Assessment of Risks Related to Wellbore Safety in Deepwater Drilling*,
https://doi.org/10.1007/978-981-95-1668-1_4

This provides a basis for the assessment of wellbore structure and engineering risks during the drilling process. The advantages of this technology can be fully utilized, combined with the theory and ideas of the pre-drilling wellbore structure risk assessment method based on probability and statistics, to seek to establish a new method for quantitatively evaluating the wellbore structure and engineering risks during deepwater drilling using logging while drilling data. Therefore, it is necessary to address the characteristics of deepwater drilling, fully utilize the advantages of technologies such as logging while drilling to seek methods to improve the prediction accuracy of formation pressure in narrow safety density windows, eliminate potential risks caused by uncertain formation parameters at the source, and on this basis, study the identification methods of engineering risks such as leakage, blowout, collapse, and sticking during the drilling process, and establish a dynamic wellbore structure risk assessment method during the drilling process. At the same time, study the uncertain parameters in the existing wellbore structure design methods and improve the design methods to minimize the risks of pre-drilling wellbore structure design.

4.1 Method for Predicting Formation Pressure with Credibility Based on Logging While Drilling Data

In order to fully consider the uncertainty of formation parameters in deepwater drilling, Guan Zhichuan et al. proposed a formation pressure prediction method based on probability and statistics. This method uses logging data or seismic data of adjacent wells to predict the formation pressure of the target well with credibility before drilling, which improves the prediction accuracy. At the same time, the pressure profile contains probability information and has the function of risk control. We can start from the principle of this method, use seismic data to predict the formation pressure of the target deepwater well with credibility, and combine the logging while drilling data to correct the pressure of the drilled formation, and then re-predict the pressure of the undrilled formation below.

4.1.1 Method for Predicting Formation Pressure with Credibility Before Drilling Based on Seismic Interval Velocity

First, introduce the method for predicting formation pressure with credibility before drilling based on seismic interval velocity.

Using seismic interval velocity to predict the pressure of the target well has a very important advantage over using logging data of adjacent wells, that is, it is not affected by the abundance of adjacent well data. At the same time, it is beneficial to correct the pressure of the drilled formation and related calculation parameters in

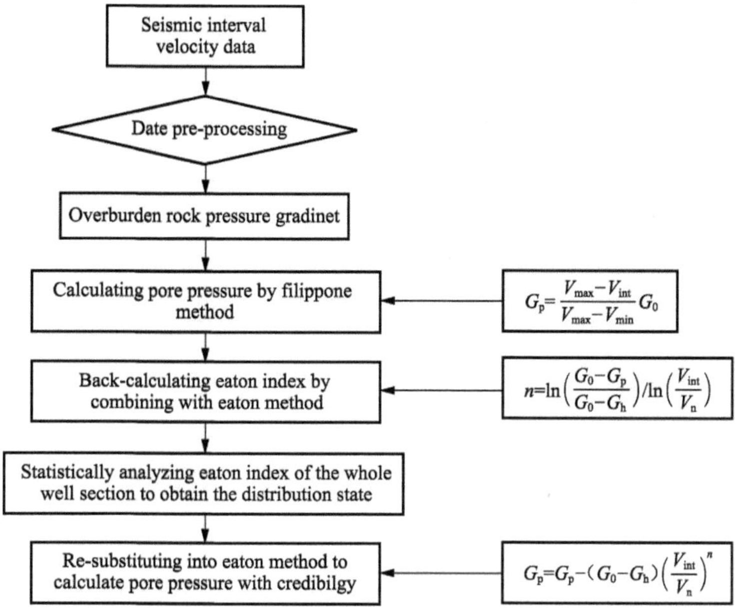

Fig. 4.1 Calculation flowchart of formation pore pressure with credibility based on seismic interval velocity

combination with logging while drilling data such as MWD (Measurement While Drilling) and LWD (Logging While Drilling).

This method uses the Eaton method, the Fillippone method, and the effective stress method to jointly calculate the formation pore pressure, and its specific process is shown in Fig. 4.1.

Using this method, the formation pore pressure distribution zone with credibility can be obtained, as shown in Fig. 4.2.

Among them, the lower limit of uncertainty is composed of the calculated values of the formation pore pressure with a cumulative probability density of 5%, and the upper limit of uncertainty is composed of the calculated values of the formation pore pressure with a cumulative probability density of 95%. Taking the calculation of the formation pore pressure at a depth of 2800 m as an example, the principle is shown in Fig. 4.3.

The Huang Rongzun method is used to calculate the formation fracture pressure, and the calculation formula is as follows:

$$P_f = \frac{3\sigma_h - \sigma_H - \left(\alpha\frac{2-3\mu}{1-\mu} - f\right)P_p + S_t}{1 - \alpha\frac{1-2\mu}{1-\mu} + f} \tag{4.1}$$

where, p_f is the formation fracture pressure, MPa; σ_H and σ_h are the maximum and minimum horizontal in-situ stresses, MPa; f is the porosity of the formation, %;

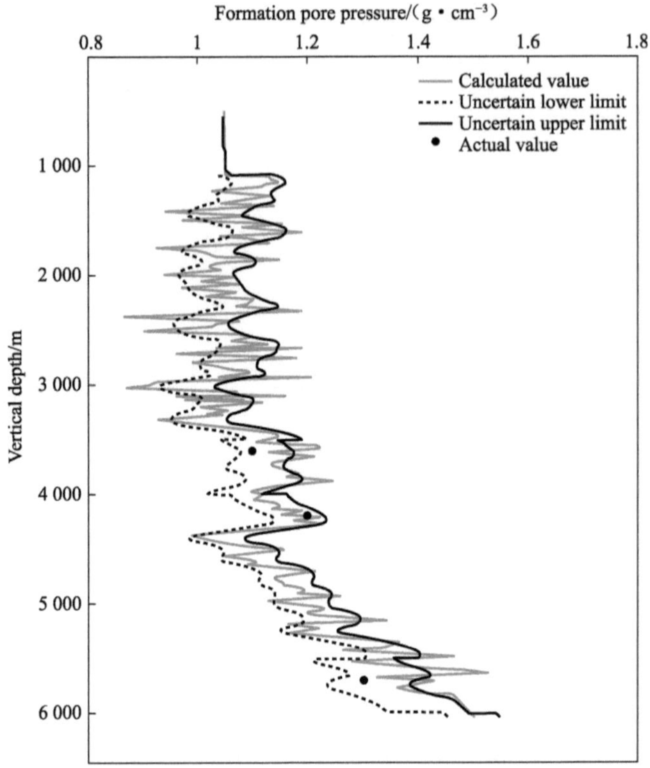

Fig. 4.2 The distribution zone of formation pore pressure with credibility

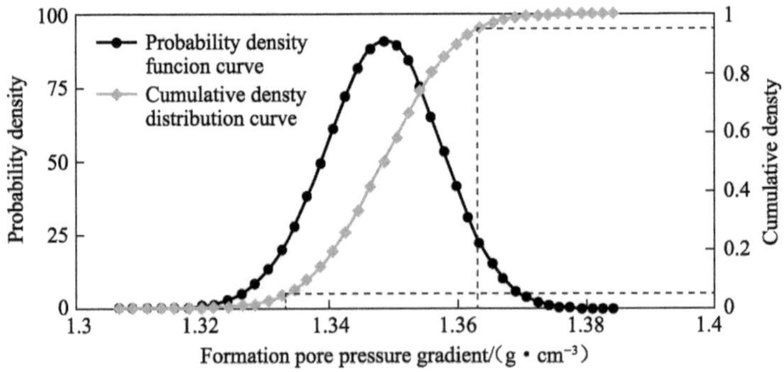

Fig. 4.3 The probability distribution of formation pore pressure at 2800 m

p_p is the formation pore pressure, MPa; S_t is the tensile strength of the formation, MPa; μ is the Poisson's ratio, dimensionless; α is the effective stress coefficient, dimensionless.

Since the calculation of the formation fracture pressure in Eq. (4.1) requires the use of the formation pore pressure, the calculated value of the formation fracture pressure is also a distribution zone with the same credibility. At the same time, the rock mechanical parameters in the formula can be calculated by transplanting the logging data of adjacent wells according to different geological layers.

4.1.2 Real-Time Correction Method of Formation Pressure with Credibility Based on Logging While Drilling Data

By analyzing the method for predicting formation pressure with credibility before drilling based on seismic interval velocity in 4.1.1, it can be known that in the flowchart of Fig. 4.1, the distribution form of the Eaton index obtained by inversion using Eq. (4.2) is the statistical result of the entire well section, as shown in Eq. (4.3) and Fig. 4.4.

$$n = \ln\left(\frac{G_0 - G_p}{G_0 - G_h}\right) / \ln\left(\frac{V_{int}}{V_n}\right) \tag{4.2}$$

$$\begin{cases} p(x) = \frac{1}{\sqrt{2\pi}\sigma} e^{-\frac{(x-\mu)^2}{2\sigma^2}} \\ F(x) = \frac{1}{\sqrt{2\pi}\sigma} \int_{-\infty}^{x} e^{-\frac{(t-\mu)^2}{2\sigma^2}} dt \end{cases} \tag{4.3}$$

Since the Eaton index changes continuously with the well depth, the method in 4.1.1 conducts probability statistics on the Eaton index of the entire well section to obtain the formation pressure distribution under different credibility conditions. Although the calculation results have sufficient accuracy, for wells where the Eaton

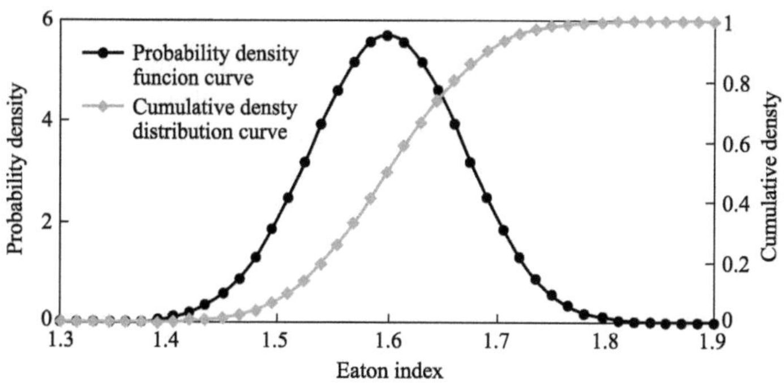

Fig. 4.4 The form distribution of Eaton index in whole section

index varies greatly along the well depth, it will result in a relatively wide distribution zone of the predicted formation pressure with credibility, which is not conducive to the wellbore structure design and risk avoidance of formations with narrow safety density windows. Therefore, the logging while drilling data can be used to re-invert and calculate the Eaton index of the drilled formation, and a reasonable method can be sought to correct the predicted pressure distribution zone of the undrilled formation below.

It is assumed that the Eaton index obtained by inversion using Eq. (4.2) based on seismic interval velocity before drilling is a large sample database $n(h)$, where h is the well depth of the target well, m; the depth of the drilled well section is h_1, and the formation pore pressure obtained from logging while drilling is G_p', $\mathrm{g/cm^3}$. Then, the Eaton index of the drilled formation can be corrected according to Eq. (4.4):

$$n'(h_1) = \ln\left(\frac{G_0 - G_p'}{G_0 - G_h}\right) / \ln\left(\frac{V_{in}}{V_n}\right) \tag{4.4}$$

where, G_h is the hydrostatic pressure, $\mathrm{g/cm^3}$; G_0 is the overburden pressure gradient, $\mathrm{g/cm^3}$; V_n is the interval velocity under normal compaction conditions, m/s; V_{in} is the interval velocity, m/s.

Regard the Eaton index of the drilled formation corrected according to the logging while drilling data as a small sample database $n'(h_1)$. According to the previous assumption, the Eaton index of the entire well section obtained by inversion based on seismic interval velocity before drilling is a large sample database $n(h)$, in which the Eaton index of the well section from 0 to h_1 (the drilled well section) is the small sample database $n(h_1)$. Replace $n(h_1)$ with $n'(h_1)$ to form a new large sample database $n'(h_1)$, and the distribution form of the Eaton index in this new large sample database can be obtained from Eq. (4.5).

$$\begin{cases} p'(x) = \frac{1}{\sqrt{2\pi}\sigma'} e^{-\frac{(x-\mu')^2}{2\sigma^2}} \\ F'(x) = \frac{1}{\sqrt{2\pi}\sigma'} \int_{-\infty}^{x} e^{-\frac{(t-\mu')^2}{2\sigma^2}} dt \end{cases} \tag{4.5}$$

Thus, according to the corrected distribution form of the Eaton index, combined with the method in 4.1.1, the distribution zone of the formation pore pressure with credibility of the undrilled formation below can be corrected in real time. The specific process is shown in the following figure (Fig. 4.5)

For the formation fracture pressure, as can be seen from Eq. (4.1), it can be corrected in real time as the formation pore pressure is corrected. At the same time, the calculation model involves many rock mechanical parameters, which are related to geological stratification and the like. In previous calculations, they were generally predicted by regional transplantation using the logging data of adjacent wells, and were regarded as fixed values within the same geological stratification for calculation. Therefore, there is uncertainty in their values. In order to obtain more accurate

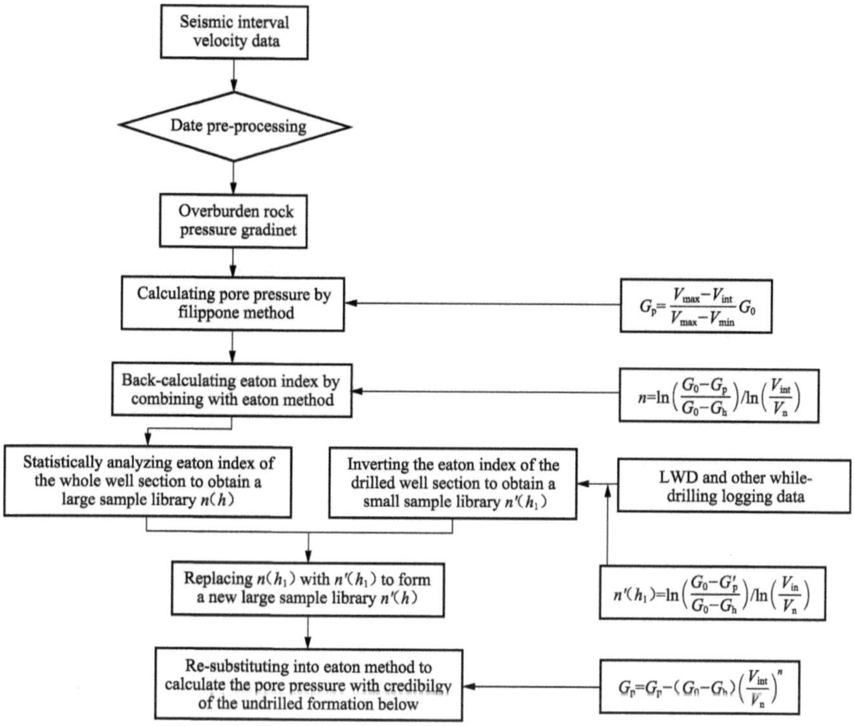

Fig. 4.5 Calculation flowchart of formation pore pressure with credibility based on LWD data

predicted values of the formation fracture pressure, the logging while drilling data can be used for correction, and the specific steps are as follows:

(1) According to the real-time corrected value of the formation pore pressure, use Eq. (4.1) to calculate the formation fracture pressure P_f' corresponding to the formation pore pressure P_p' at different probabilities under the credibility condition J, as shown in Eq. (4.6):

$$P_f' = \frac{3\sigma_h - \sigma_H - \left(\alpha\frac{2-3\mu}{1-\mu} - f\right)P_p' + S_t}{1 - \alpha\frac{1-2\mu}{1-\mu} + f} \tag{4.6}$$

(2) If the rock mechanical parameters $x(\sigma_h, \sigma_H, \alpha, \mu, f, S_t)$ of the next geological stratification predicted by using the data of adjacent wells before drilling are known, when drilling to a distance of Δh (the step size is set according to the actual situation) from the top of the next geological stratification, it is considered that the average value $\overline{x'}\left(\overline{\sigma_h'}, \overline{\sigma_H'}, \overline{\alpha'}, \overline{\mu'}, \overline{f'}, \overline{S_t'}\right)$ of each rock mechanical parameter $x'\left(\sigma_h', \sigma_H', \alpha', \mu', f', S_t'\right)$ within this geological stratification obtained from logging while drilling is the true value of this stratification. Use Eq. (4.7)

to recalculate the formation fracture pressure P_{f}'' of this geological stratification:

$$P_{\mathrm{f}}'' = \frac{3\overline{\sigma_{\mathrm{h}}'} - \overline{\sigma_{\mathrm{H}}'} - \left(\overline{\alpha'}\frac{2-3\overline{\mu'}}{1-\overline{\mu'}} - \overline{f'}\right)P_{\mathrm{p}}' + \overline{S_{\mathrm{t}}'}}{1 - \overline{\alpha'}\frac{1-2\overline{\mu'}}{1-\overline{\mu'}} + \overline{f'}} \tag{4.7}$$

(3) If drilling continues for δh, and the average value of each rock mechanical parameter within the well section of $\Delta h + \delta h$ of this geological stratification measured by logging while drilling is $\overline{x''}$, when $\left|\overline{x''} - \overline{x'}\right| > \varepsilon$ (ε is a set allowable error), then replace $\overline{x'}\left(\overline{\sigma_{\mathrm{h}}'}, \overline{\sigma_{\mathrm{H}}'}, \overline{\alpha'}, \overline{\mu'}, \overline{f'}, \overline{S_{\mathrm{t}}'}\right)$ in Eq. (4.7) with $\overline{x''}$ to recalculate the formation fracture pressure P_{f}''' of this geological stratification;
(4) Repeat step (3) until the current geological stratification is drilled through.

Thus, a method for predicting formation pressure with credibility based on logging while drilling data is established. Using this method, the formation pore pressure and formation fracture pressure with credibility of the undrilled formation below can be corrected in real time based on the logging while drilling data. It not only has the advantages of the method for predicting formation pressure with credibility before drilling in 4.1.1, but also can make up for the prediction errors caused by the insufficient abundance of adjacent well data, thereby improving the pressure prediction accuracy of the undrilled formation below.

Taking a certain offshore drilling abroad as an example, this well is a drilled well, and its logging data and adjacent well data are relatively comprehensive. The water depth of this well is 58 m, and the total depth of the well is 4267 m.

First, use the method in 4.1.1 to predict the formation pressure profile of the entire well section with credibility (as shown in Fig. 4.6a). Suppose that when drilling to 1520 m, the logging data of the drilled well section can be used as the logging while drilling data, and use the method established in this paper to re-predict the formation pressure of the undrilled formation below, as shown in Fig. 4.6b.

As can be seen from Fig. 4.6, the formation pressure distribution zone corrected by using the logging while drilling data becomes narrower. Thus, on the premise that the credibility remains unchanged, the prediction accuracy of the formation pressure with credibility of the undrilled formation below is improved, which is beneficial to the wellbore structure design and engineering risk avoidance in deepwater areas with narrow safety density windows.

With the application of intelligent algorithms such as machine learning in the field of drilling geomechanics, the methods for real-time updating of formation pressure while drilling have been further developed. The author has also proposed a prediction and while-drilling correction method for formation pressure with confidence levels based on well-seismic data fusion, which can integrate drilling, logging, and seismic data from more regions to quantitatively characterize and update the uncertainties in formation pressure.

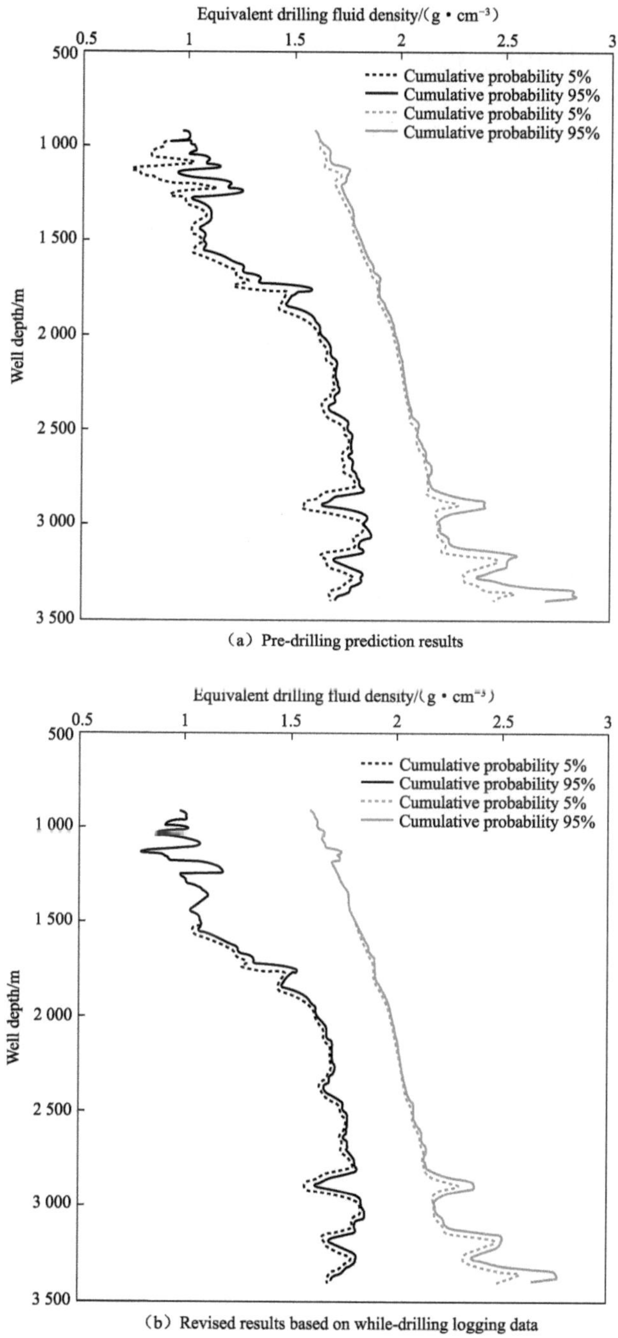

Fig. 4.6 Formation pore pressure prediction with credibility based on LWD data

4.2 Dynamic Wellbore Structure Risk Assessment Method During Drilling Process

4.2.1 Determination of the Safe Drilling Fluid Density Window with Credibility

The method for predicting formation pressure with credibility based on logging while drilling data established in 4.1.2 can correct the predicted values of the formation pressure with credibility of the undrilled formation below in real time as the drilling depth increases. Therefore, the safe drilling fluid density window of the undrilled formation below is also corrected in real time accordingly.

Since the formation pressure profile established in 4.1 is a distribution zone with credibility, if the credibility $J = 95\% - 5\% = 90\%$ is selected, then the formation pore pressure profile corresponding to the cumulative probability of 95% is defined as the lower limit curve of the safe drilling fluid density window, and the formation fracture pressure profile corresponding to the cumulative probability of 5% is defined as the upper limit curve of the safe drilling fluid density window. The safe drilling fluid density window with credibility J is formed, as shown in Fig. 4.7.

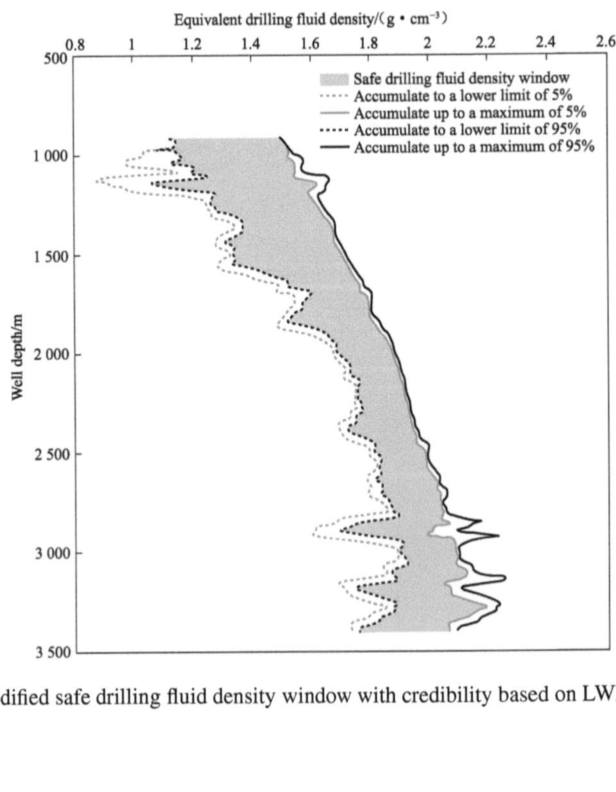

Fig. 4.7 Modified safe drilling fluid density window with credibility based on LWD data

Within the safe drilling fluid density window in Fig. 4.7, it is considered that the prediction of the formation pressure has a credibility of J, that is, under the condition of credibility J, this safe drilling fluid density window has sufficient accuracy.

4.2.2 Dynamic Wellbore Structure Risk Assessment

Within the safe drilling fluid density window with credibility corrected based on the logging while drilling data established in 4.2.1, the risk assessment of the existing wellbore structure design scheme can be carried out according to the pressure balance constraint conditions.

In order to ensure the safety of the open hole section, the following pressure balance constraint conditions need to be met.

Anti-kick when stopping the pump:

$$\rho_{\mathrm{m}}(H_i) \geq \rho_{\mathrm{pmax}} + S_{\mathrm{b}} + \Delta\rho \tag{4.8}$$

Anti-pipe sticking due to pressure difference:

$$\left(\rho_{\mathrm{m}}(H_i) - \rho_{\mathrm{pmin}}\right) \times H_{\mathrm{pmin}} \times 0.00981 \leq \Delta P \tag{4.9}$$

Anti-leakage when running in:

$$\rho_{\mathrm{m}}(H_i) + S_{\mathrm{g}} + S_{\mathrm{f}} \leq \rho_{\mathrm{fmin}} \tag{4.10}$$

Anti-leakage when shutting in the well:

$$\rho_{\mathrm{m}}(H_i) + S_{\mathrm{k}}\frac{H_i}{H_{\mathrm{n}-1}} + S_{\mathrm{f}} \leq \rho_{\mathrm{f}}(H_{\mathrm{n}-1}) \tag{4.11}$$

where, H_i is the depth of the calculation point, m; S_{b} is the swab pressure coefficient, g/cm^3; S_{k} is the kick allowance, g/cm^3; S_{g} is the surge pressure coefficient, g/cm^3; $\rho_{\mathrm{m}}(H_i)$ is the static equivalent density of the drilling fluid, g/cm^3; ρ_{pmax} is the maximum value of the formation pore pressure (the lower limit curve of the safe drilling fluid density window) of the formation at and above the calculation point, g/cm^3; ρ_{pmin} is the minimum value of the formation pore pressure (the lower limit curve of the safe drilling fluid density window) of the formation at and above the calculation point, g/cm^3; Δp represents the allowable value of sticking due to pressure difference, MPa; S_{f} is the safety coefficient of the formation fracture pressure, g/cm^3; $\Delta\rho$ is the additional drilling fluid density, g/cm^3; ρ_{fmin} is the minimum value of the formation fracture pressure (the upper limit curve of the safe drilling fluid density window) of the well section at and above the calculation point, g/cm^3; H_{pmin} is the well depth where the minimum formation pore pressure is located, m; $H_{\mathrm{n}-1}$ is the depth of the casing shoe of the upper casing string, m.

Since both the kick allowance S_k and the safety coefficient of the formation fracture pressure S_f are safety coefficients artificially added due to the inaccurate prediction of the formation pressure, which represent the uncertainty of the formation pressure. And within the safe drilling fluid density window with credibility J, it is considered that the prediction of the formation pressure is accurate and reliable enough. Therefore, at this time, the kick allowance S_k and the safety coefficient of the formation fracture pressure S_f can be ignored, and then Eq. (4.11) in the pressure balance constraint conditions can be ignored, and Eq. (4.10) becomes:

Anti-leakage when running in:

$$\rho_m(H_i) + S_g \le \rho_{fmin} \tag{4.12}$$

Thus, Eqs. (4.8), (4.9) and (4.12) constitute the pressure balance constraint conditions under the condition of credibility J.

If the designed drilling fluid density of the undrilled formation below in the existing wellbore structure design scheme satisfies Eqs. (4.8), (4.9) and (4.12), it is considered that there are no engineering risks such as leakage, blowout, collapse and sticking in the current wellbore structure design scheme. If not, the risk $R_{Kick,StickingduetoPressureDifference,LeakageduringDrilling(h)}$ at the calculated well depth h can be quantitatively evaluated according to the following formula.

Kick risk:

$$R_{K(h)} = p(\rho_m(h) - S_b - \Delta\rho < \rho_{min}(h)) = 1 - F_{\rho_p(h)}(\rho_m(h) - S_b - \Delta\rho) \tag{4.13}$$

Sticking due to pressure difference risk:

$$R_{S(h)} = p\left(\rho_m(h) - \frac{\Delta P}{0.00981 \cdot h} > \rho_{min}(h)\right) = F_{\rho_p(h)}\left(\rho_m(h) - \frac{\Delta P}{0.00981 \cdot h}\right) \tag{4.14}$$

Leakage risk during drilling:

$$R_{L(h)} = p\big(\rho_m(h) + S_g > \rho_{max}(h)\big) = F_{\rho_f(h)}\big(\rho_m(h_i) + S_g\big) \tag{4.15}$$

where, $R_{K(h)}$, $R_{S(h)}$, $R_{L(h)}$ are the magnitudes of the kick, sticking due to pressure difference, and leakage risks during drilling at the well depth h, respectively, %; $p(x)$ is the probability magnitude corresponding to the event x; $\rho_{max}(h)$ and $\rho_{min}(h)$ are the equivalent densities corresponding to the lower limit curve and the upper limit curve of the safe drilling fluid density window with credibility at the well depth h, respectively, g/cm^3; $F_{\rho_p(h)}$ is the cumulative probability distribution of the formation pore pressure at the well depth h; $F_{\rho_f(h)}$ is the cumulative probability distribution of the formation fracture pressure at the well depth h.

In the wellbore structure risk assessment model of the undrilled formation below composed of Eqs. (4.13), (4.14), and (4.15), $\rho_{max}(h)$ and $\rho_{min}(h)$ are updated in real time according to the logging while drilling data using the method in 4.1. Therefore, the dynamic risk assessment of the existing wellbore structure design scheme can be carried out during the drilling process. If the evaluation result indicates a high engineering risk, the existing wellbore structure design scheme should be adjusted, and the risk assessment should be carried out again until the pressure balance constraint conditions under the condition of credibility J are met.

4.2.3 Example Calculation

Taking the offshore drilling example in 4.1.2 as an example, this well is a drilled well, and its logging data and downhole accidents such as leakage, blowout, collapse, and sticking (wellbore structure risks) are all known. It can be used as a reference to verify the accuracy and reliability of this method. When drilling to 1520 m, the formation pressure with credibility of the undrilled formation below, which is corrected by using the logging while drilling data (logging data of the drilled section), is shown in Fig. 4.6b. The engineering risks of the existing wellbore structure design scheme of the lower formation (see Table 4.1 and Fig. 4.8) are evaluated by using the method in this section. The statistical situation of the actual drilling engineering accidents (wellbore structure risks) is shown in Fig. 4.8, and the wellbore structure risk assessment results of this method are shown in Fig. 4.9.

In Fig. 4.9, it is clearly shown that there is a kick risk at around 1700 m of the well depth, and the risk probability is close to 100%; there are leakage risks at around 1800 m and 2600 m of the well depth, and the magnitudes of these risks are close to 100%. By comparing with the statistical results of the actual drilling accidents in Fig. 4.8, it can be seen that in the calculation results above 3000 m of the well depth, several engineering risks with higher risks are consistent with the accident types and the corresponding well depths recorded in the actual drilling, which indicates that this method can effectively conduct a quantitative assessment of the wellbore structure risks for a quite long well section of the undrilled formation below, and the assessment results have high accuracy. At the same time, as the drilling depth gradually increases, the correction by using the logging while drilling data further improves the prediction accuracy of the formation pressure of the undrilled formation below. Furthermore, the

Table 4.1 The current casing program design data

Casing stage	Hole	MD/ft	Casing	MD/ft	MD/m	Drilling fluid/ppg
Conductor pipe	36	525	30	507	155	8.7
First spud	26	3020	20	3005	916	8.7–14
Second spud	17	5997	13-(3/8)	5955	1815	9.5–15.8
Third spud	12-(1/4)	8088	9-(5/8)	8060	2457	13.4–15.8
Fourth spud	8-(1/2)	11,030	7	11,028	3361	15.3–16.7
Fifth spud	6	14,000	6	14,000	4267	14.7–15

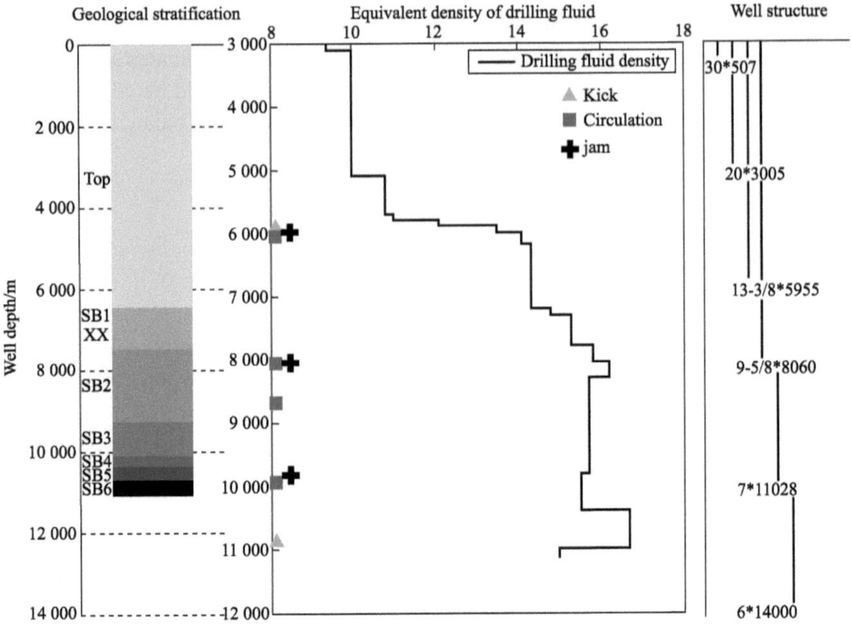

Fig. 4.8 The statistical results of project risk in actual drilling

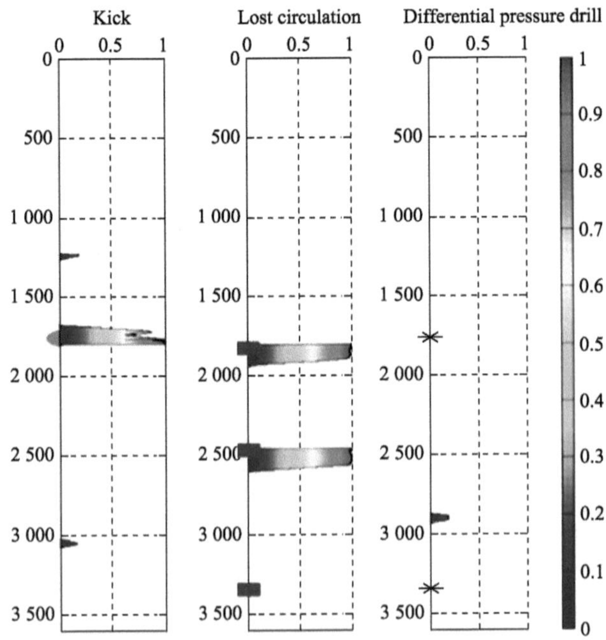

Fig. 4.9 Dynamic risk assessment results of casing program design

accuracy and reliability of the assessment results of the existing wellbore structure risks of the lower formation by using this method will also be further improved.

4.3 Improvement of Wellbore Structure Design Method

In order to minimize the wellbore structure risk at the design stage and make full use of the narrow safety density window in deepwater drilling, it is necessary to improve the existing wellbore structure design method.

4.3.1 Improvement of Pre-Drilling Wellbore Structure Design Method

The top-down design method is mostly used in deepwater drilling. This can ensure that each layer of casing is run to the deepest depth, reserve enough casing levels for the lower well section, and has achieved good results. However, the constraint conditions of the safety density window in this method still use the traditional calculation method, which is difficult to meet the requirements of the design accuracy for the narrow safety density window in deepwater drilling. It will cause waste of the already narrow density window, and may even lead to inaccurate design results or difficulty in designing to the target formation, increasing the risk of downhole complex accidents. Therefore, it is necessary to carry out a more refined wellbore structure design for the problem of the narrow safety density window in deepwater drilling, make full use of the safety density window, improve the accuracy of the wellbore structure design results, and minimize the risk of pre-drilling wellbore structure design.

4.3.1.1 Analysis of Traditional Safety Density Window Constraint Conditions and Their Limitations

The safety density window constraint conditions in the top-down wellbore structure design method are based on the pressure profiles (formation pore pressure profile and fracture pressure profile) of the formation in the area to be drilled, as well as the drilling data and wellbore structure design coefficients in this area, so that the open-hole section satisfies the mechanical equilibrium relationship composed of Eqs. (4.8), (4.9), (4.10), and (4.11).

Then Eqs. (4.8) and (4.11) constitute the constraint conditions of the safety density window in the traditional top-down wellbore structure design. Through analysis, it is not difficult to find that the equivalent density of the drilling fluid in the constraint conditions is the static equivalent density, resulting in the constraint conditions ignoring the annular pressure loss caused by drilling circulation, that is, lacking the anti-leakage constraint condition during drilling:

Anti-leakage during drilling:

$$\rho_{\mathrm{m}}(H_i) + S_{\mathrm{u}} + S_{\mathrm{f}} \leq \rho_{\mathrm{f}min} \tag{4.16}$$

where, S_{u} is the equivalent density of the annular pressure loss, $\mathrm{g/cm^3}$.

According to the mechanical equilibrium relationship, to ensure that there is no well leakage during the circulation drilling process, the equivalent density of the drilling fluid must satisfy the above formula.

According to the design principle of the casing level and setting depth, the equivalent density of the drilling fluid in the wellbore needs to be kept within the safety window, and this principle needs to be satisfied during the entire drilling construction process (i.e., tripping, circulation drilling, well shut-in, etc.). However, when circulating and drilling, the equivalent density of the drilling fluid at the bottom of the wellbore will increase by an equivalent density of the annular pressure loss S_{u} on the basis of the original static equivalent density. If the well depth is relatively deep, that is, S_{u} is relatively large, the equivalent density of the drilling fluid designed according to Eqs. (4.8)–(4.11) will be too large, resulting in well leakage during the circulation drilling process.

Now carry out a specific analysis: For wells with a relatively shallow conventional well depth, the equivalent density of the drilling fluid designed by the conventional method is safe, mainly for two reasons: First, the coefficients such as S_{b}, S_{k}, S_{g}, S_{f} in the conventional design are all taken as relatively large values; Second, the annular pressure loss is not very large, and the constraint condition (4.11) makes up for the deficiency of not considering the annular pressure loss in Eq. (4.10). The former, although increasing the safety of the design result, wastes the safety density window. For formations with a narrow safety density window, it may lead to the inability to complete the design. By comparing Eq. (4.10) with Eq. (4.11), it can be seen that the difference between the two lies in S_{g} and $S_{\mathrm{k}} \times \frac{H_i}{H_{\mathrm{n}-1}}$, as well as ρ_{fmin} and $\rho_{\mathrm{f}}(H_{\mathrm{n}-1})$. For conventional formations, ρ_{fmin} and $\rho_{\mathrm{f}}(H_{n-1})$ are equal or not very different, and the surge pressure coefficient S_{g} generally takes values between $0.01-0.05\mathrm{g/cm^3}$, and the kick allowance S_{k} generally takes values between $0.05-0.10\mathrm{g/cm^3}$. Then $S_{\mathrm{k}} \times \frac{H_i}{H_{\mathrm{n}-1}}$ generally takes values between 0.0625 and 0.125 (taking the open-hole section of 2000–2500 m as an example). At the same time, the equivalent density of the annular pressure loss at this time is approximately between $0.03-0.05\mathrm{g/cm^3}$. It can be seen that the equivalent density of the drilling fluid obtained from the constraint condition (4.11) is less than the result obtained from the constraint condition (4.16) considering the equivalent density of the annular pressure loss S_{u}. Therefore, there will be no well leakage during the circulation drilling process.

However, for conventional deep wells, ultra-deep wells, and deepwater drilling, the equivalent density of the annular pressure loss S_{u} increases. In order to ensure that there is no well leakage during the circulation drilling process, S_{u} may be greater than $S_{\mathrm{k}} \times \frac{H_i}{H_{\mathrm{n}-1}}$ in the constraint condition (4.11). Therefore, the constraint condition (4.16) needs to be considered. Take a well with a well depth of 6500 m as an example.

The depth of the casing shoe in the open-hole section is 5200 m. At this time, the equivalent density of the annular pressure loss S_u reaches 0.07g/cm^3, its surge pressure coefficient S_g is 0.03g/cm^3, and the kick allowance S_k is 0.05g/cm^3. Then $S_k \times \frac{H_i}{H_{n-1}} = 0.0625 < S_u = 0.07$. It can be seen that if the anti-leakage constraint condition (4.16) during drilling is not considered, the designed equivalent density of the drilling fluid may be greater than the actual safe value, resulting in well leakage during the circulation drilling process.

Through the above analysis, the safety density window constraint conditions of the conventional top-down wellbore structure design method, due to the use of multiple empirical safety coefficients and taking relatively large values to ensure the safety of the design results, to a certain extent, cause waste of the safety density window, and even lead to the inability to reach the target formation. For deep wells and deepwater wells, the equivalent density of the drilling fluid designed by the constraint conditions of the traditional method is larger than the actual safe equivalent density, which will lead to the risk of well leakage during the circulation drilling period. Therefore, for formations with a narrow safety density window in deep wells and deepwater drilling, it is necessary to improve the constraint conditions of the safety density window, make full use of the narrow safety density window, and establish a more refined wellbore structure design method.

4.3.1.2 Pre-drilling Wellbore Structure Design Method Considering ECD and Precise Calculation of Design Coefficients

As can be seen from the previous analysis, for deep wells and deepwater wells, especially those wells with a narrow safety density window in the lower formation, it is necessary to consider the influence of the annular pressure loss on the safety density window. At the same time, the empirical safety coefficients in the constraint conditions of the safety density window should be improved, and the upper and lower limits of the safety density window should be accurately determined, so as to make full use of the originally narrow safety density window for wellbore structure design.

(1) Precise Determination of the Upper and Lower Limits of the Safe Drilling Fluid Density Window

Introduce the pre-drilling formation pressure prediction method with credibility (the method introduced in 4.1.1 or the prediction method based on the logging data of adjacent wells) to predict the formation pressure profile, obtain the formation pressure distribution zone containing probability information, and determine the safe drilling fluid density window under the condition of credibility J according to the method in 2.1. On this basis, Eqs. (4.8), (4.9) and (4.12) form the new pressure balance constraint conditions under the condition of credibility J, in which Eqs. (4.9) and (4.12) jointly determine the upper limit of the safety density window.

For the lower limit of the safety density window, by analyzing the new constraint conditions, only Eq. (4.8) is the lower limit constraint condition of the safety density window. Its main parameters are: ρ_{pmax}, S_b and $\Delta\rho$, among which ρ_{pmax} can be accurately determined under a certain credibility condition by using the formation pressure prediction method with credibility; and the additional drilling fluid density $\Delta\rho$ is mainly related to factors such as the prediction accuracy of the formation pore pressure, the burial depth of the oil, gas and water layers, the hydrogen sulfide content in the oil and gas, and the matching situation of the well control devices, and it is an empirical coefficient that needs to be comprehensively considered. First, according to the regional drilling data, use the method introduced in 1.3.2.1 of Chap. 1 for transplantation prediction to obtain the predicted value $\Delta\rho^*$ containing probability information at the target well. Then, according to the matching situation of the well control devices and the selection situation of the formation pressure prediction credibility, comprehensively determine the reasonable value range of $\Delta\rho$. Since the pressure profile with credibility is introduced, within this credibility range, it can be considered that the prediction of the formation pore pressure is accurate enough. Therefore, $\Delta\rho$ can take the minimum value within the predicted value range.

The swab pressure coefficient S_b is mainly related to factors such as the wellbore structure, the bottom hole assembly, the performance of the drilling fluid (density, flow behavior index and consistency coefficient), and the tripping speed. Through investigation and calculation: The swab pressure coefficient increases with the increase of the tripping speed, roughly showing a linear distribution; it increases with the increase of the consistency coefficient, roughly showing a linear distribution, and the variation range is larger compared with the tripping speed; it increases with the increase of the flow behavior index, roughly showing an exponential trend; while the variation range with the density of the drilling fluid is very small, and the influence of the density of the drilling fluid on it can be ignored; at the same time, the influence range of the wellbore structure and the bottom hole assembly on the swab pressure coefficient is within $0.02\,\text{g/cm}^3$, and the areas with greater influence are mainly concentrated when the well depth is small, and the influence of the wellbore structure can be ignored when the well depth is large.

For the upper formation, since the safety density window is relatively wide, the specific influence of the wellbore structure and the bottom hole assembly on the swab pressure coefficient can be ignored and a relatively large value can be selected; for the lower formation, especially the formation with a narrow safety density window, since the influence of the wellbore structure on the swab pressure coefficient is very small at this time, only a typical wellbore structure needs to be selected as the basis, and factors such as the tripping speed, the flow behavior index and the consistency coefficient of the drilling fluid need to be comprehensively considered for design. In this way, the swab pressure coefficient of the lower formation calculated in this way generally takes values between $0.01-0.05\,\text{g/cm}^3$, avoiding the waste of the narrow safety density window of the lower formation.

(2) Introduction of Annular Pressure Loss into the Constraint Conditions

The annular pressure loss is a calculated value related to factors such as the wellbore structure, the rheology of the drilling fluid and the displacement, and it can only be calculated when a given wellbore structure is available, thus limiting its introduction, which is also the main reason why the conventional safety density window constraint

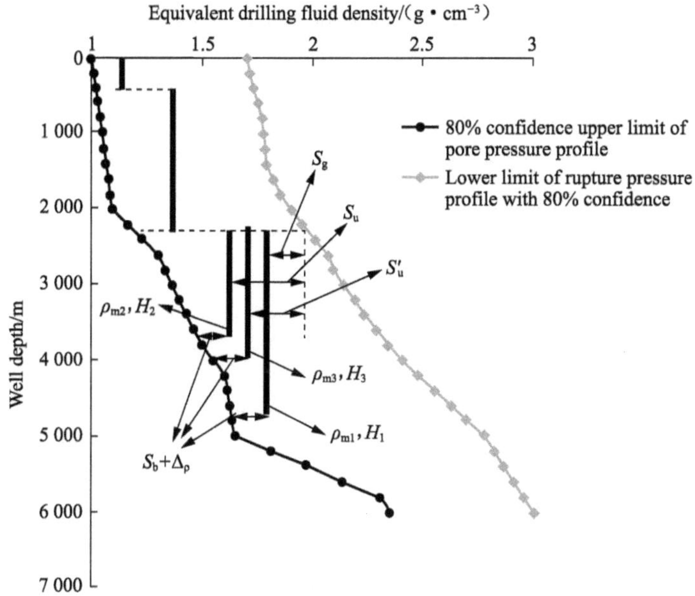

Fig. 4.10 Schematic diagram for iterative calculation of annulus pressure loss equivalent density

conditions lack the annular pressure loss. In order to solve this problem, iterative calculation can be adopted.

In Fig. 4.10, the formation pore pressure profile and the formation fracture pressure profile are respectively the upper limit curve of the pore pressure profile with 80% credibility and the lower limit curve of the fracture pressure profile with 80% credibility obtained by the method in 4.1.1. Suppose the setting depth design of the third layer of casing is carried out, and the specific design steps are as follows.

1 First, design the equivalent density of the drilling fluid ρ_{m1}, and the setting depth H_1 according to the newly established safety density window constraint conditions (4.8), (4.9), and (4.12) previously.
2 Calculate the equivalent density of the annular pressure loss S_u at this time according to ρ_{m1}, H_1, the parameters of the upper wellbore structure, the performance parameters of the drilling fluid, the displacement, etc.
3 Replace S_g in the constraint condition (4.12) with S_u, then the constraint condition (4.12) becomes:

$$\rho_m(H_i) + S_u \leq \rho_{fmin} \tag{4.17}$$

4 Use the method in step ①, design the equivalent density of the drilling fluid ρ_{m2} and the setting depth H_2 according to the constraint conditions (4.8), (4.9), and (4.17), and repeat step ② to calculate the equivalent density of the annular pressure loss S'_u at this time.

5 Repeat step ③, then the constraint condition (4.17) becomes:

$$\rho_{\mathrm{m}}(H_{\mathrm{i}}) + S'_{\mathrm{u}} \leq \rho_{\mathrm{fmin}} \quad (4.18)$$

6 Finally, design the equivalent density of the drilling fluid ρ_{m3} and the setting depth H_3 according to the constraint conditions (4.8), (4.9), and (4.18). After two iterations, ρ_{m3} and H_3 are considered as the designed density of the drilling fluid and the setting depth of the casing.

Through the above steps, the annular pressure loss related to the wellbore structure can be introduced into the pre-drilling wellbore structure design. Generally, sufficient accuracy can be achieved after two iterations. If higher requirements are needed, the iterative solution can be continued according to the same method until a satisfactory equivalent density of the drilling fluid and the setting depth of the casing are obtained.

So far, a pre-drilling wellbore structure design method has been established based on the formation pressure profile with credibility, taking Eqs. (4.8), (4.9), (4.12), and (4.17) as the constraint conditions of the safety density window and considering the annular pressure loss. This method not only makes full use of the narrow safety density window of the lower formation, but also avoids well leakage during circulation drilling by adding ECD, and improves the reliability of the pre-drilling wellbore structure design results.

4.3.1.3 Example Calculation

Take a certain deep well as an example. The designed well depth is 6500 m. This well is designed by using the conventional top-down wellbore structure design method. Its formation pressure profile and the design results of the wellbore structure are shown in Fig. 4.11. Before the fifth spudding, when the pressure test is carried out with the designed drilling fluid density, the upper casing shoe (at a depth of 5200 m) is safe. However, during the drilling process, when drilling to 5220 m, a well leakage accident occurs, which brings troubles to the drilling operation.

The design result of this well reflects that the density of the drilling fluid calculated by using the conventional safety density window constraint conditions is on the high side. For the lower formation, especially the formation with a narrow safety density window, well leakage is likely to occur during circulation drilling.

Now, the method proposed in this paper is used to redesign this well, and the design result is shown in Fig. 4.12.

In Fig. 4.12, due to the introduction of the formation pressure profile with 80% credibility, the density window is slightly narrower than that in Fig. 4.10. However, as a result, the kick allowance and the safety coefficient of the fracture pressure in the constraint conditions are cancelled. At the same time, by accurately calculating the additional drilling fluid density and the swab pressure, the setting depth of each layer of casing is made deeper, which is beneficial to the safety of deep wells and deep exploration wells where the lower formation is not fully understood. Moreover,

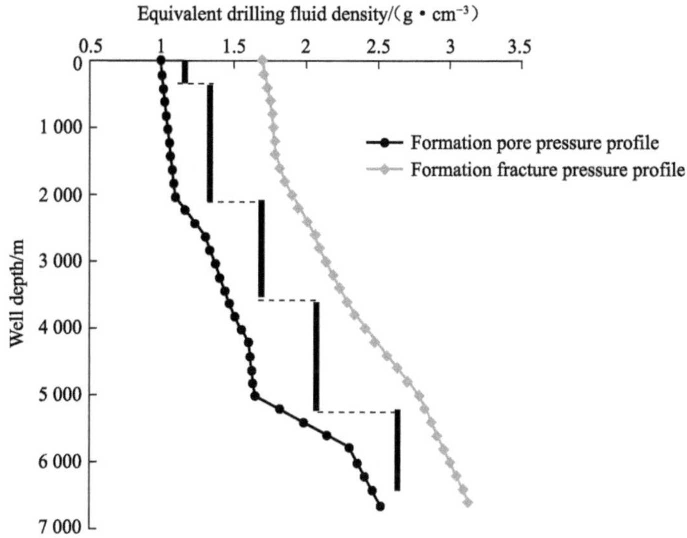

Fig. 4.11 Formation pressure profile and casing program design results

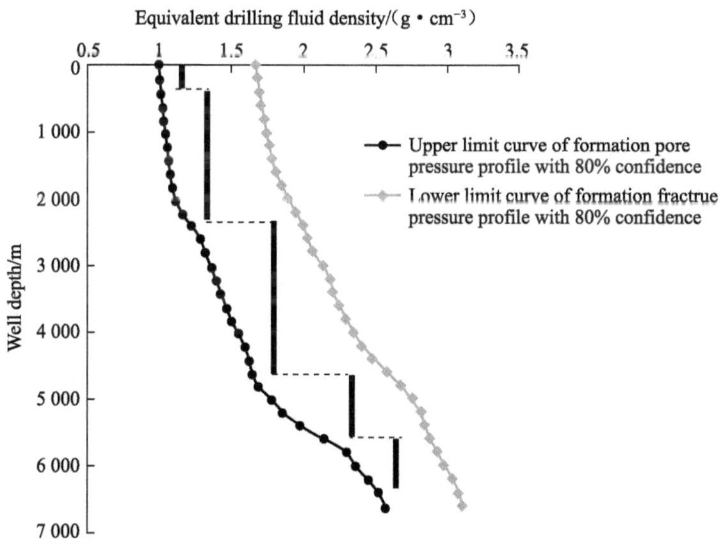

Fig. 4.12 Formation pressure profile with credibility and precise casing program design results

the equivalent density of the annular pressure loss is introduced in this design result, avoiding the occurrence of well leakage accidents during the circulation drilling process.

The comparison of the casing levels and the designed setting depths is shown in Table 4.2.

Table 4.2 Comparison of design results

Casing stage	Running depth (m)	
	Traditional design method	Refined design method
Surface casing	450	450
Technical casing 1	2155	2368
Technical casing 2	3786	4753
Technical casing 3	5200	5689
Production casing	6500	6500

It is worth noting that the author subsequently proposed the concept of a wellbore ECD (Equivalent Circulating Density) matrix, which is used to provide a more detailed and scientific description of wellbore pressure changes in the open hole section during drilling before drilling operations. This method has been applied to the optimized design of wellbore structures for two 10,000-m scientific exploration wells in China.

4.3.2 Dynamic Wellbore Structure Design Method During Drilling Process

A pre-drilling wellbore structure design method has been established previously, which is based on the pre-drilling formation pressure prediction profile with credibility, takes Eqs. (4.8), (4.9), (4.12) and (4.17) as the constraint conditions of the safety density window, and considers the annular pressure loss. This method improves the reliability of the pre-drilling wellbore structure design results, thereby reducing the potential risks of the pre-drilling wellbore structure design scheme.

However, through the analysis in 4.1, although the pre-drilling formation pressure prediction method with credibility can improve the reliability of formation pressure prediction, for deepwater drilling with low abundance of adjacent well data, it is still necessary to use logging while drilling data to correct it in real time, so as to maximize the prediction accuracy of the formation pressure. At the same time, if the method introduced in 4.2 predicts that there are relatively large risks in the existing wellbore structure design scheme of the undrilled formation below during the drilling process, it should be adjusted to avoid risks. Therefore, this paper proposes a dynamic wellbore structure design method during the drilling process based on logging while drilling data, and its specific process is shown in Fig. 4.13.

Among them, the new constraint conditions of the safe drilling fluid density window are as follows:

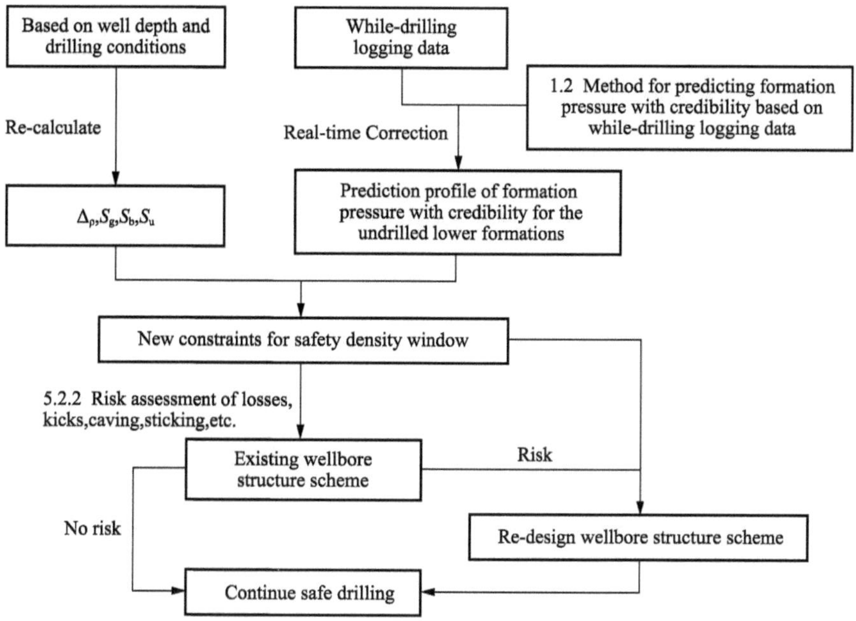

Fig. 4.13 Flow chart of dynamic risk assessment of casing program design in drilling process

$$\begin{cases} \text{Prevent well kick when stopping the pump}: \rho_m(H_i) \geq \rho_{pmax'} + S_b' + \Delta\rho' \\ \text{Prevent casing sticking due to pressure difference}: \left(\rho_m(H_i) - \rho_{pmin}'\right) \times H_{pmin} \\ \times 0.00981 \leq \Delta p \\ \text{Prevent leakage when running in the drill string}: \rho_m(H_i) + S_g' \leq \rho_{fmin}' \\ \text{Prevent leakage during drilling}: \rho_m(H_i) + S_u' \leq \rho_{fmin}' \end{cases} \tag{4.19}$$

where, ρ_{pmax}', ρ_{pmin}', ρ_{fmin}', S_b', S_g', S_u', $\Delta\rho'$ are no longer fixed values, but variables that change with the drilling process. ρ_{pmax}', ρ_{pmin}', ρ_{fmin}' are respectively the maximum formation pore pressure gradient, the minimum formation pore pressure gradient and the minimum formation fracture pressure gradient within the predicted formation pressure profile of the undrilled formation below after being corrected based on the measurement data while drilling, g/cm^3; S_b', S_g', S_u', $\Delta\rho'$ are respectively the swab pressure coefficient, the surge pressure coefficient, the equivalent density of the circulation pressure loss and the additional drilling fluid density calculated based on the real-time wellbore structure and drilling working conditions, g/cm^3.

This method uses the logging while drilling data to correct the formation pressure profile with credibility of the undrilled formation below in real time. At the same time, it calculates the equivalent density of the circulation pressure loss S_u, the additional density of the drilling fluid $\Delta\rho$, the surge pressure coefficient S_g and the swab pressure coefficient S_b in real time according to the drilling depth and working conditions, so as to determine the new constraint conditions of the safety density window (Eq. (4.19)). On this basis, the method in 2.2 is used to conduct a quantitative assessment of the accident risks such as leakage, blowout, collapse and sticking of the existing wellbore structure of the undrilled formation below. Once a higher risk is found, the wellbore structure design scheme of the lower formation is redesigned in a timely manner, so as to carry out real-time dynamic wellbore structure design for the undrilled formation below during the drilling process and avoid engineering risks to the greatest extent at the design level.

4.4 Summary

(1) A method for predicting formation pressure with credibility based on logging while drilling data has been established. This method can correct the pressure distribution zone with credibility of the undrilled formation below in real time according to the logging while drilling data, improving the pressure prediction accuracy of the undrilled formation below.

(2) Based on the method for predicting formation pressure with credibility based on logging while drilling data, combined with the constraint conditions of the safety density window and probability theory, a dynamic wellbore structure risk assessment method during the drilling process has been established. It can conduct real-time quantitative assessment of the wellbore structure risks of the undrilled formation below during the drilling process.

(3) By analyzing the constraint conditions of the safety density window in the existing wellbore structure design method, it is found that the values of the empirical safety coefficients are too large and the prevention of leakage during circulation drilling is not considered, which is unfavorable for formations with a narrow safety density window in deepwater areas. Therefore, a pre-drilling refined wellbore structure design method considering ECD and precise calculation of design coefficients has been established; at the same time, a dynamic wellbore structure design method during the drilling process has been proposed. It can carry out real-time dynamic wellbore structure design for the undrilled formation below during the drilling process, thus avoiding engineering risks to the greatest extent at the design level.

(4) With the continuous deepening of drilling depth, the formations faced by drilling become increasingly complex and variable. As a safety barrier to isolate complex formations during drilling, the contradiction of insufficient casing levels in wellbore structure has become more prominent. Scientific, refined, and customized optimized design of wellbore structure is becoming increasingly important. Of course, with the application and development of artificial intelligence, especially large language models, intelligent optimization decision-making technologies for wellbore structure based on big data and large models will soon emerge.

Bibliography

1. Guangming Z, Shihong W, Wenyuan H (2012) Hotspot trend and enlightenment of global ten-year hydrocarbon exploration [J]. Acta Petrolei Sinica 33(S1):14–19
2. Yuhong X (2015) Status and prospect of proprietary oil and gas field exploration and development in deepwater West of South China Sea [J]. Oil Drilling Prod Technol 37(1):5–7
3. Ke K, Zhichuan G, Junya Z et al (2009) Affecting factors research on the risks of deep water drilling operation in West Africa [J]. Oil Drilling Prod Technol 31(5):5–10
4. Zhiliang Q, Shiguo W, Zhijun W et al (2011) Geohazards and risk of deepwater engineering induced by gas hydrate—a case study from oil leakage of deepwater drilling well in GOM [J]. Progress Geophys 26(4):1279–1287
5. Shaughnessy JM, Daugherty WT, Graff RL et al (2007) More ultradeepwater drilling problems [C]. SPE 105792
6. Caihong L, Jin Y, Shijing C et al (2008) Drilling riser mechanical characteristic of marine deep water [J]. Oil Drilling Prod Technol 2:28–31
7. Peng W, Tie Y, Shanshan L et al (2014) A coupled mechanical model of deep water riser and the drill string [J]. J Oil Gas Technol 36(6):83–86, 6
8. Xiuquan L, Guoming C, Yuanjiang C et al (2013) Analyses and countermeasures of deepwater drilling riser grounding accidents under typhoon conditions [J]. Pet Explor Dev 40(6):738–742
9. Zhichuan G, Kanhua S, Yinao S (2009) Analysis on lateral load-bearing capacity of conductor and surface casing for deepwater drilling [J]. Acta Petrolei Sinica 30(2):285–290
10. Zhichuan G (2003) Effect of temperature and pressure on fluid column pressure of wellbore drilling fluid in deep-water drilling with oil-based drilling fluid [J]. J Univ Petrol China (4):48–52, 57–149
11. Yuanjiang C, Guoming C, Liangbin X et al (2009) Influential factors for the design of ultra-deepwater drilling risers [J]. Petrol Exploration Dev 36(4):523–528
12. Youhua W, Wenhai W, Xingxun J (2011) South China Sea deepwater drilling challenges and solutions [J]. Petrol Drilling Techn 39(2):50–55
13. Zhi Y, Honghai F, Guobin Z et al (2010) Investigation of shallow water flow in deepwater drilling [J]. Petrol Drilling Techn 38(6):48–52
14. Jin Y, Bailing Z, Bo Z et al (2015) Geological disaster acoustic wave identification and prediction technology of deepwater shallow gas [J]. Oil Drilling Prod Technol 37(1):143–146
15. Zhiyuan W, Baojiang S, Haiqing C et al (2008) Prediction of gas hydrate formation region in the wellbore of deepwater drilling [J]. Pet Explor Dev 35(6):731–735
16. Stahl B, Baur MP (1983) Design methodology for offshore platform conductors [J]. J Petrol Technol 35:12(12):1973–1984
17. Da Costa DFO, Rodrigues RS, Negrao AF (1990) Evolution of deepwater drilling in Brazil [C]. SPE 21158
18. King GW (1990) Drilling engineering for subsea development wells [J]. J Petrol Technol 42(9):1176–1183

Y. Xu, *Assessment of Risks Related to Wellbore Safety in Deepwater Drilling*,
https://doi.org/10.1007/978-981-95-1668-1

19. Barltrop N, Adams A (1991). Dynamics of fixed marine structures [M]. Butterworth-Heinemann
20. Wetmore SB, Halbleib BL Consideration for design of drilling conductors for the new generation of deepwater, harsh weather jackups [C]. SPE 23856
21. King GW, Kevin D, Trevor H (1993) A coupled analysis approach to the assessment of marine drilling systems [J]. SPE Drill Complet 8(2):131–137
22. Thorogood JL, Train AS, Adams AJ (1998) Deep water riser system design and management [C]. SPE/IADC 39295
23. Rocha LAS, Junqueira P, Roque JL (2003) Overcoming deep and ultra deepwater drilling challenges [C]. SPE/OTC 15233
24. Watson P, Kolstad E, Borstmayer R et al (2003) An innovative approach to development drilling in deepwater Gulf of Mexico[C]. SPE/IADC 79809
25. Juiniti R, Salies J, Polillo A (2003) Campos Basin: lessons learned and critical issues to be overcome in drilling and completion operations [C]. SPE/OTC 15221
26. Shaughnessy JM, Daugherty WT, Graff RL et al (2007) More ultra-deepwater drilling problems [C]. SPE 105792
27. Evans J, McGrail J (2011) An evaluation of the fatigue performance of subsea wellhead systems and recommendations for fatigue enhancements [C]. SPE/OTC 21400
28. Sunday K, Ward P, Griffin C, Lorenson P (2015) Detailed approach for the assessment of accumulated wellhead fatigue [C]. SPE 175441
29. Howells H, Baker R, Rimmer A (2015) Measurement of wellhead fatigue [C]. SPE/OTC 25684
30. Chunming S (2000) Stability analysis of subsea wellhead systems [D]. SouthWest Petroleum University
31. Dawei G (2003) Analysis of marine riser failure [J]. Oil Field Equip 5:4–7
32. Kanhua SU, Zhichuan G, Yinao S (2008) Mechanical stability analysis of subsea wellhead for deepwater drilling [J]. Oil Drilling Prod Technol 30(6):1–4, 15
33. Kanhua SU (2009) Analysis on mechanical stability of subsea wellhead and bearing capacity of conductor for deepwater drilling [D]. Chinese Doctoral Dissertations Full-text Database
34. Li W (2009) Study on deepwater drilling wellhead stability analysis of West Africa JDZ-2-1 Well [D]. Chinese Master's Theses Full-text Database
35. Kanhua S, Zhichuan G, Lu W et al (2009) Analysis on subsea wellhead stability of surface BOP drilling system in deepwater operations [J]. China Offshore Oil Gas 21(3):180–185
36. Hui Z, Deli G, Haixiong T et al (2010) Tubular mechanics in jetting operation of conductor under deepwater condition [J]. Acta Petrolei Sinica 31(3):516–520
37. Yi W, Jin Y, Chun L et al (2011) The ANSYS-based analysis of the mechanical performance of underwater wellhead [J]. China Petrol Mach 39(12):35–37, 99
38. Kanhua SU (2012) Review of coupled analysis between the casing string and the seabed soil for deepwater drilling [J]. J Chongqing Univ Sci Technol (Nat Sci Edn) 14(3):43–46
39. Kanhua S, Zhichuan G, Zhihui L (2012) Impact of the seabed soil properties on the stability of subsea wellhead for deepwater drilling [J]. Oil Drilling Prod Technol 34(3):7–10, 23
40. Shaodong J (2012) Research on operation analysis and decision for deepwater drilling riser and subsea wellhead [D]. Chinese Doctoral Dissertations Full-text Database
41. Jiwei L (2012) The research of bearing capacity of conductor for deepwater drilling [D]. Chinese Master's Theses Full-text Database
42. Qingzhi Y (2014) Research on bearing capacity at time of jetted conductor in deepwater after installation [D]. Chinese Master's Theses Full-text Database
43. Yanbin W, Deli G, Jun F (2014) Numerical analysis of bearing capacity of deep water conductor with consideration of different contract interface models between pile and soil [J]. China Offshore Oil Gas 26(5):76–82
44. Duns H Jr, Ros NCJ (1963) Vertical flow of gas and liquid mixtures in wells[a]. Sixth world petroleum congress: section2 [C], Paper No.22, Frankfurt
45. Zuber N, Findlay JA (1965) Average volumetric concentration in two-phase flow systems [J]. Trans ASME 87(2):453–468

46. Orkiszewski J. (1967) Predicting two-phase pressure drops in vertical pipe [J]. J.P.T. 19(6): 829–838
47. LeBlanc JL, Lewis RL (1968) A Mathematical model of a gas kick [J]. J Petrol Technol SPE 1860, 20(8):889–890
48. Taitel Y, Bomea D, Dukler AE (1980) Modeling flow pattern transitions for steady upward gas-liquid flow in vertical tubes [J]. AICHE 26(3):345–354
49. Nickens HV (1987) A dynamic computer model of a kicking well [J]. SPE Drilling Eng SPE 14183, 2(2):159–173
50. Hasan AR, Kabir CS (1988) A study of multiphase flow behavior in vertical wells [J]. SPEPE 5:263–274
51. Santos OLA (1991) Well-control operation in horizontal wells [J]. SPE Drilling Eng SPE 21105, 6(2):111–117
52. Ansari AM, Sylvester ND et al (1994) A comprehensive mechanistic model for upward two-phase flow in wellbores [J]. SPE Prod. & Faci. 9(2):143–151
53. Ohara S (1995) Improved method for selecting kick tolerance during deepwater drilling operations[R]. Louisiana State University, Baton Rouge
54. Choe J, Juvkam-Wold HC (1997) A modified two-phase well control model and its computer applications as a training and educational tool [J]. SPE Comput Appl SPE 37688, 12(4):14–20
55. Gabor T (2001) Considerations on the selection of an optimum vertical multiphase pressure drop prediction model for oil wells [C]. SPE 68361
56. Nunes JO , Bannwart AC, Ribeiro PR (2002) Mathematical modeling of gas kicks in deep water scenario [C]. SPE 77253
57. Jun F, Guang C et al (2000) An underbalanced drilling theoretical model and its application [J]. Acta Petrolei Sinica (4):75–79, 123
58. Xiangfang L, Xiang qi Z, Xiuxiang S et al (2004) Study on two-phase gas-liquid flow during gas kick [J]. J Eng Thermophys (1):73–76
59. Yingcao Z, Deli G, Yonggui L (2005) New model for calculating bottom hole pressure of multiphase flow in annulus of underbalanced straight well [J]. Acta Petrolei Sinica 02:96–99
60. Zhiyuan W, Baojiang S, Yonghai G et al (2008) Simulation computation of well killing with deepwater driller's method [J]. Acta Petrolei Sinica 5:786–790
61. Baojiang S, Zhiyuan W, Peibin G et al (2011) Application of a seven-component multiphase flow model to deepwater well control [J]. Acta Petrolei Sinica 32(06):1042–1049
62. Yonghai G, Baojiang S, Xinxin Z et al (2012) Multiphase flow in wellbores and variation laws of the bottom hole pressure in gas hydrate drilling [J]. Acta Petrolei Sinica 33(5):881–886
63. Lianwei S, Zhichuan G, Haiming L et al (2013) Effects of casing program change and liner tieback on well control [J]. Petroleum Drilling Techniques 41(6):34–39
64. Meipeng R, Xiangfang L, Qingtao M et al (2013) A new design method of killing fluid density against blowout during tripping [J]. Petrol Drilling Techn 41(1):25–30
65. Gensheng L, Liangbin D, Shouceng T et al (2013) Characteristics of wellbore transient flow during sour gas influx [J]. Petrol Drilling Techn 41(4):8–14
66. Shihui S (2014) Research on wellbore flow model for managed pressure drilling at constant Bottomhole Pressure [D]. Chinese doctoral dissertations full-text database
67. Chaoyang X, Yingfeng M, Na W et al (2015) Experimental simulation and numerical modeling of dynamic variations in wellbore pressure during gas-kicks [J]. Acta Petrolei Sinica 36(1):120–126
68. Zhichuan G, Chunshan L, Guangchen Z et al (2001) A method for designing casing program in deep and superdeep wells [J]. J Univ Petrol China (6):42–44, 6
69. Zhiming Y, Guoming C, Liangbin X et al (2006) Optimization of casing program of deepwater well by dual-gradient drilling [J]. Nat Gas Indus (12):112–114, 204–205
70. Zhichuan G, Ke K, Baoping L (2009) An approach to casing program design with formation pressure uncertainties [J]. J China Univ Petrol (Edn Nat Sci) 33(4):71–75
71. Bo T (2009) Determination of safe window of drilling fluid density for deepwater drilling [D]. Chinese Master's Theses Full-text Database

72. Zhichuan G, Ke K, Kanhua S (2011) Casing design and optimization for deepwater drilling [J]. Petrol Drilling Techn 39(2):16–21
73. Guoming C, Xiuquan L, Yuanjiang C et al (2013) Advances in technology of deepwater drilling riser and wellhead [J]. J China Univ Petrol (Edn Nat Sci) 37(5):129–139
74. Guojin Y, Yi G, Qinman T, Wilson JF (1991) Structural dynamics of the ocean [M]. Beijing: Petroleum Industry Press
75. Zhong T, Linhai H, Yongqian Z et al (2004) Flexural behaviors of concrete-filled double-skin (CHS inner and SHS outer) steel tubes [J]. Indus Constr (1):6–9, 5
76. Zhong T, Linhai H, Yongqian Z et al (2004) Flexural behaviors of concrete-filled double-skin (CHS inner and SHS outer) steel tubes [J]. Indus Constr 34(1):6–10
77. Linhai H, Youfu Y (2004) Modern concrete-filled steel tube structure technology [M]. China Architecture & Building Press, Beijing, pp 70–71
78. Matlock H (1970) Correlations for design of laterally loaded piles in soft clay[C]. OTC 1204
79. Reese LC (1974) Analysis of laterally loaded piles in sand [C]. OTC 2080
80. Burke BG (1974) An analysis of marine risers for deep water[C]. OTC 1771
81. Shishen L, Yachao L (1987) Calculation and analysis of pile foundation [M]. Beijing: China Communications Press, 104–110
82. Zongyu Y (2005) On the selection of positive and dimensionless methods in multi-index comprehensive evaluation [J]. Stat Theory Practi (4):24–25
83. Weihua S (2000) Research on the theory and method of multi-index comprehensive evaluation [D]. Xiamen University
84. Mazmishvili (1984) Error theory and least squares method [M]. China Coal Industry Publishing House
85. China Classification Society (2005) Classification and construction code of offshore mobile platform [M]. Beijing: China Communications Press
86. Yuqiang X, Zhichuan G, Qing W et al (2015) Determination and risk assessment methods for setting depth of conductor during deepwater drilling [J]. J China Univ Petrol (Edn Nat Sci) 39(6):85–91
87. Akers TJ (2008) Jetting of structural casing in deepwater environments: job design and operational practices [J]. SPE Drill Complet 23(1):29–40
88. Hui Z, Deli G, Haixiong T (2009) Mechanics analysis of jetting string during jetting operation of conductor [J]. J Southwest Petrol Univ (Sci Technol Edn) 31(6):148–151, 218–219
89. Shunwen W, Jin Y, De Y et al (2012) Research of jetting drilling mechanism of surface conductor in deepwater [J]. J Oil Gas Technol 34(8):157–160, 169
90. Yuqiang X, Zhichuan G, Kanhua S (2014) Methods of setting depth range with credibility of conductor for deepwater drilling based on probability statistics [J]. Appl Ocean Res 48:301–307
91. Renyi Z (2005) Theory and application of spatial variation [M]. Science Press, Beijing
92. Russo D, Bouton M (1992) Statistical analysis of spatial variability in unsaturated flow parameters [J]. Water Resour Res 28(7):1911–1925
93. Ping C, Tianshou M (2014). Research status of early monitoring technology for deepwater drilling overflow [J]. Acta Petrolei Sinica 35(3):602–612
94. Chen W (2011) Status and challenges of Chinese deepwater oil and gas development [J]. Pet Sci 8(4):477–484
95. Jin Y, Shijing C (2008) Current situation and developing trend of petroleum drilling technologies in deep water [J]. Oil Drilling Prod Technol 2:10–13
96. Lubin Z, Yunhua G, Haige W (2009) Early kick detection methods for deepwater drilling and its future development [J]. Oil Drilling Prod Technol 31(1):22–26
97. Meipeng R, Xiangfang L, Darong X et al (2012) Research of kick and distribution features of gas-liquid two phase flow during drilling [J]. J Eng Thermophys 33(12):2120–2125
98. Yonghai G, Baojiang S, Zhiyuan W et al (2008) Calculation and analysis of wellbore temperature field in deepwater drilling [J]. J China Univ Petrol (Edn Nat Sci) 2:58–62
99. Liwei N (2008) The effect of drilling fluid's physical properties on the temperature and pressure in deepwater drilling [D]. Chinese Master's Theses Full-text Database

100. Xuncheng S, Zhichuan G (2011) Full transient analysis of heat transfer during drilling fluid circulation in deep-water wells [J]. Acta Petrolei Sinica 32(4):704–708
101. Yonghai G (2007) Study on multi-phase flow in wellbore and well control in deep water drilling [D]. Chinese Doctoral Dissertations Full-text Database
102. Toralde JSS, Wuest CH (2014) Riser gas risk mitigation with advanced flow detection and managed pressure drilling technologies in deepwater operations [C]. OTC 24997-MS
103. Parth P (2014) Innovative kick detection system for HP/HT ultradeepwater wells using a section of the BHA [C]. OTC 25074-MS
104. Toskey ED (2014) Kick detection at the subsea mudline[C]. OTC 25847-MS
105. Xiuxiang S, Yufeng L, Yiming L et al (2014) Early monitoring of the gas-cut in deepwater riser based on doppler measuring system [J]. Petrol Drilling Techn 42(5):90–94
106. Xingliang D, Shijing C, Haixiong T et al (2011) Offshore drilling handbook [M]. Petroleum Industry Press, Beijing
107. Tiecai L, Xizhi L (2014). Similarity and similarity principle [M]. Harbin: Harbin Institute of Technology Press
108. Liang QJ (2002) Application of quantitative risk analysis to pore pressure and fracture gradient prediction [C]. SPE 77354
109. Yuanyuan Y, Li W (2010) Types of geological hazards in ultra-deep water wells [J]. Sci Technol Ecno Market (9):35–36
110. Bo Z, Jin Y, Bailing Z et al (2012) Prediction and control technology of shallow geological hazards in deepwater area [J]. Marine Geol Frontiers 28(01):51–54
111. Jianming G, Yanqiu Y, Guijingl Y et al (2010) Research on the bottom boundary of thermogenic gas hydrate stability zone in Shenhu Area, South China Sea [J]. J China Univ Min Technol 39(6):870–875
112. Digby AJ (2005) Assessment and quantification of the hydrate geohazard [C]. OTC 17223
113. Mohammadi AH, Tohidi B (2005) A novel predictive technique for estimating the hydrate inhibition effects of single and mixed thermodynamic inhibitors [J]. Canadian J Chem Eng 83(6):951–961
114. Dickens GR, Quinby-Hunt MS (1997) Methane hydrate stability in pore water: a simple theoretical approach for geophysical applications [J]. J Geophys Res Solid Earth (1978–2012) 102(B1):773–783
115. Yuxing L, Shuchu F (1992) (2) Method of identifying gas hydrate formation in pipeline [J]. Nat Gas Indus (2):null
116. Shuhong W, Haibin S, Wen Y (2005) The change of external conditions effects on the phase equilibrium curve of gas hydrate and the thickness of hydrate stability zone [J]. Prog Geophys 3:761–768
117. Najibi H, Chapoy A, Haghighi H et al (2009) Experimental determination and prediction of methane hydrate stability in alcohols and electrolyte solutions [J]. Fluid Phase Equilib 275(2):127–131
118. Østergaard KK, Masoudi R, Tohidi B et al (2005) A general correlation for predicting the suppression of hydrate dissociation temperature in the presence of thermodynamic inhibitors [J]. J Petrol Sci Eng 48(1):70–80
119. Yousif MH, Young DB (1993) A simple correlation to predict the hydrate point suppression in drilling fluids[C] SPE/IADC 25705
120. Ouar H, Cha SB, Wildeman TR et al (1992) The formation of natural gas hydrates in water-based drilling fluids [J]. Chem Eng Res Des 70(A1):48–54
121. Yuqiang X, Zhichuan G, Chuanbin X et al (2015) Risk evaluation methods of gas hydrate formation in the wellbore of deepwater drilling [J]. Acta Petrolei Sinica, 36(5):633–640
122. Yuqiang X, Zhichuan G, Chuanbin X et al (2015) Risk evaluation methods of gas hydrate when shallow strata drilling in deepwater area [C]. OTC 25727-MS
123. Gang L, Xiaosen L (2010) Experimental investigation of the supercooling effect on methane hydrate formation [J]. Geoscience 24(3):627–631, 637
124. Sha H, Zhiming W (2010) High sour gas based on analytic hierarchy process [J]. Drilling Prod Technol 33(2):28–30, 37, 137

125. Yuqiang X, Z G hichuan, Hongning Z et al (2014) Methods for precise casing design based on formation pressure profiles with different credibility [J]. J China Univ Petrol (Edn Nat Sci) 38(6):72–77
126. Xiongwen Y, Yingcao Z, Shiliang F et al (2010) Strategy analysis of narrow window drilling technology and practice [J]. Oil Field Equip 39(8):7–11
127. Drilling Manual (Party A) Writing Group. Drilling manual (Party A) [M]. Beijing: Petroleum Industry Press (1990)
128. Tinggen C, Zhichuan G (2000) Theory and technology of drilling engineering [M]. China University of Petroleum Press, Dongying, pp 251–254
129. Honghai F, Yuanlin C, Xisheng L (1995) Prediction for wellbore dynamic surge pressure while tripping a drillpipe [J]. J Univ Petrol China 1995(5):null
130. Zhichuan G, Xuncheng S (1996) Determination of annulus clearance between casing string and borehole under the condition of limited surge pressure [J]. J Univ Petrol China (6):null
131. Honghai F, Xisheng L (1990) Analysis on surge pressure caused by mud viscosity in a vertical well [J]. J Univ Petrol China 2:8–15
132. Qin S, Xutian H (2011) Research on drilling design techniques for narrow mud weight window [J]. Petrol Drilling Techn 39(3):62–65
133. Yuqiang X, Zhichuan G, Huizeng Z et al (2016) The quantitative description of gas-cut degree in deepwater drilling [J]. Petrol Explor Dev 43(2):292–296
134. Pan L, Jianchun F, Shujie L (2014) Quantitative analysis of borehole collapse in drilling base on fault tree analysis and Bayesian networks [J]. J Safety Sci Technol 10(1):143–149.
135. Shujie L, Xiangfang L, Yue Z et al (2014) Reliability analysis on safety barriers of deep-water drilling based on Bayesian-LOPA method [J]. J Safety Sci Technol 10(9):187–191

MIX
Papier aus verantwortungsvollen Quellen
Paper from responsible sources
FSC® C105338

If you have any concerns about our products,
you can contact us on
ProductSafety@springernature.com

In case Publisher is established outside the EU,
the EU authorized representative is:
Springer Nature Customer Service Center GmbH
Europaplatz 3, 69115 Heidelberg, Germany

Printed by Libri Plureos GmbH
in Hamburg, Germany